Lecture Notes in Mathematics

Edited by A. Dold and B. Eckmann

Subseries: Fondazione C.I.M.E., Firenze
Adviser: Roberto Conti

1224

Nonlinear Diffusion Problems

Lectures given at the 2nd 1985 Session of the Centro Internazionale Matematico Estivo (C.I.M.E.) held at Montecatini Terme, Italy June 10 – June 18, 1985

Edited by A. Fasano and M. Primicerio

Springer-Verlag
Berlin Heidelberg New York London Paris Tokyo

Editors

Antonio Fasano
Mario Primicerio
Istituto Matematico Università, Viale Morgagni 67/A
50134 Firenze, Italy

Mathematics Subject Classification (1980): 35-02, 35 A 25, 35 B 05, 35 B 32, 35 F 20, 35 J 55, 35 K 65, 35 P 15, 35 R 35, 58 E 07, 58 G 10, 76 S 05, 80 A 25, 92 A 15

ISBN 3-540-17192-4 Springer-Verlag Berlin Heidelberg New York
ISBN 0-387-17192-4 Springer-Verlag New York Berlin Heidelberg

© Springer-Verlag Berlin Heidelberg 1986
Printed in Germany

Printing and binding: Druckhaus Beltz, Hemsbach/Bergstr.
2146/3140-543210

PREFACE

This volume contains the texts of the three series of lectures given at the C.I.M.E. Session on "Some Problems in Nonlinear Diffusion" held at "La Querceta", Montecatini, from June 10 to June 18, 1985.

The general theme of the session was the study of the effects of nonlinearity in diffusion problems. Two main topics were considered: diffusion problems with degeneracy (such as in the porous media equation), and reaction-diffusion problems.

The first topic has been treated in the lectures by prof. Donald G. Aronson (University of Minnesota, Minneapolis). He considered a variety of aspects, ranging from physical background to regularity and asymptotic behaviour of solutions, also including peculiar subjects like waiting times and Hamilton-Jacobi equation.

Mathematical modelling of reaction-diffusion problem with reference to the chemical engineering applications has been illustrated by prof. Ivar Stakgold (University of Delaware, Newark). Various types of approximations have been discussed and the corresponding mathematical aspects have been investigated devoting special attention to the possible formation of dead cores.

Steady state processes in reaction-diffusion have been the main subject of the lectures by prof. Jesus Hernandez (Universidad Autonoma, Madrid). His large overview of qualitative methods covers in particular comparison arguments, the stability of solutions, and the use of topological degree theory.

The volume is complemented by a seminar on "rearrangements of functions and partial differential equations" which was presented by prof. Giorgio Talenti (Università di Firenze).

We wish to thank the lecturers and the participants, as well as the CIME scientific committee. We feel that the Session was quite successful for the interest shown by the audience and the extremely high quality of the lectures delivered.

A. Fasano

M. Primicerio

TABLE OF CONTENTS

C.I.M.E. Session on "Some Problems in Nonlinear Diffusion"

List of Participants

O. ARENA, Istituto Matematico Università, Viale Morgagni 67/A, 50134 Firenze

D.G. ARONSON, School of Mathematics, 127 Vincent Hall, 206 Church Street, Minneapolis, Minnesota 55455, USA

M. BARDI, Seminario Matematico Università, Via Belzoni 7, 35131 Padova

E. BERETTA, Via P. Lumumba 3, 61029 Urbino

L. BOCCARDO, Istituto Matematico Università, Città Universitaria, 00185 Roma

V. CAPASSO, Dipartimento di Matematica, Campus Universitario, 70125 Bari

E. COMPARINI, Istituto Matematico Università, Viale Morgagni 67/A, 50134 Firenze

R. DAL PASSO, Istituto per le Applicazioni del Calcolo, Viale del Policlinico 137, 00161 Roma

Z. DIVIS, Ohio State University, Department of Mathematics, 231 W. 18th Ave., Columbus, Ohio 43210, USA

A. FASANO, Istituto Matematico Università, Viale Morgagni 67/A, 50134 Firenze

P. FISCON, Via Tommaso da Celano 22, 00179 Roma

Z. GUAN, Department of Mathematics, Hangzhou University, China

J. HERNANDEZ, Departamento de Matematicas, Universidad Autonoma, Madrid, Spain

A. JERNQVIST, Doktor Forselius gata 34, S-413 26 Goteborg, Sweden

P. KNABNER, University of Augsburg, Institute for Mathematics, Memminger Str. 6, D-8900 Augsburg, West Germany

P. KUMLIN, Department of Mathematics, CTH, S-412 96 Goteborg, Sweden

M.R. LAYDI, 15 rue des Deux Princesses, 25 Besançon, France

H. LENNERSTAD, Chalmers University of Technology, Department of Mathematics, S-412 96 Goteborg, Sweden

P. MARCATI, Istituto Matematico Università, Via Roma 33, 67100 L'Aquila

A. MENDEZ ALONSO, Ronda de Segovia n. 22, 28005 Madrid, Spain

J.-C. MIELLOU, Domaine de Château, Devecey, 25870 Geneuille, France

E. MIRENGHI, Via della Reistenza 48/B, 70125 Bari

M. NIEVES GARCIA GARCIA, José del Hierro 4, 28027 Madrid

R.H. NOCHETTO, Istituto di Analisi Numerica del CNR, Corso Carlo Alberto 5, 27100 Pavia

G. PAPI FROSALI, Istituto Matematico Università, Viale Morgagni 67/A, 50134 Firenze

E. PESSA, Via Salaria per l'Aquila 72, 02100 Rieti

M.A. POZIO, Dipartimento di Matematica, II Università di Roma, Via O. Raimondo, 00173 Roma

M. PRIMICERIO, Istituto Matematico Università, Viale Morgagni 67/A, 50134 Firenze

A. PUGLIESE, Dipartimento di Matematica, Università, 38050 Povo, Trento

R. RICCI, Istituto Matematico Università, Viale Morgagni 67/A, 50134 Firenze

R. SALVI, Dipartimento di Matematica del Politecnico, Via Bonardi 9, 20133 Milano

M. SHU, Department of Mathematics, Sichuan University, Chengdu, Sichuan, China

E. SOCOLOVSKY, Mathematics Dept. U-9, Univ. of Connecticut, Storrs, CT 06268, USA

I. STAKGOLD, Department of Mathematical Sciences, University of Delaware, 501 Ewing Hall, Newark, Delaware 19716, USA

J. SWETINA, Institut fur Theoretische Chemie und Strahlenchemie, Wahringerstrasse 17, A-1090 Wien, Austria

G. TALENTI, Istituto Matematico Università, Viale Morgagni 67/A, 50134 Firenze

D.A. TARZIA, Pasaje Espora 61, 2000 Rosario, Argentina

L. TUBARO, Dipartimento di Matematica, Università, 38050 Povo, Trento

M. UGHI, Istituto Matematico Università, Viale Morgagni 67/A, 50134 Firenze

B. VENTURI, Istituto Matematico Università, Viale Morgagni 67/A, 50134 Firenze

C. VERDI, Dipartimento di Matematica, Università, Corso Carlo Alberto, 27100 Pavia

S.M. VERDUYN LUNEL, Centrum voor Wiskunde en Informatica, Kruislaan 413, 1098 SJ Amsterdam, The Netherlands

THE POROUS MEDIUM EQUATION

D.G. Aronson
School of Mathematics
University of Minnesota
Minneapolis, MN 55455/USA

Introduction

These lectures are intended as a brief introduction to the mathematical study of nonlinear diffusion mechanisms. Rather than attempt a comprehensive survey, I have elected to focus my attention on the so called porous medium equation. This allows me to convey the broad outlines of the theory without too many technicalities. Even with this restriction, it is impossible in eight lectures to cover everything of importance that has been done in the past twelve or so years. The choices I have made reflect my personal taste and should not be taken as a value judgement on the omitted material. Because of the volume of material covered, the level of proof is quite variable. Some results are presented without proof, some with fairly detailed sketches of proof, and still others with only cryptic hints. In any event, references to the main sources are always provided. Various time constraints have made it impossible for me to make an extensive revision of my original lecture notes. Thus the text presented here, though lacking in polish, is very close to what was actually said in the lectures.

Outline

- Lecture 1. Physical background. Selfsimilar solutions. Basic existence and uniqueness theory.
- Lecture 2. Basic estimates. Regularity results for one dimensional flows.
- Lecture 3. The interface in one dimensional flow: waiting times, local smoothness, corner points, asymptotic behavior, ultimate smoothness.
- Lecture 4. The porous medium equation as a finite speed approximation to a Hamilton-Jacobi equation.
- Lecture 5. Regularity in d-dimensional flow: global Holder continuity, ultimate Lipschitz continuity, local counterexamples.
- Lecture 6. Initial trace theory.
- Lecture 7. Asymptotic behavior of solutions to initial value and boundary value problems in $\mathbb{R}^d$.
- Lecture 8. Stabilization theory.

Lecture 1

PHYSICAL BACKGROUND. Consider an ideal gas flowing isentropically in a homogeneous porous medium. The flow is governed by the following three laws [M].

Equation of state: $p = p_0 \rho^{\alpha}$,
where $p = p(x,t)$ is the pressure, $\rho = \rho(x,t)$ is the density, and $\alpha \in [1,\infty)$ and $p_0 \in \mathbb{R}^+$ are constants. Here $x \in \mathbb{R}^d$ for some $d \geq 1$.

Conservation of mass: $\varkappa \frac{\partial \rho}{\partial t} + \operatorname{div}(\rho \vec{v}) = 0$,
where $\vec{v} = \vec{v}(x,t)$ is the velocity vector and $\varkappa \in \mathbb{R}^+$ is the porosity of the medium (i.e., the volume fraction available to the gas).

Darcy's Law: $\nu \vec{v} = -\mu \nabla p$,
where $\nu \in \mathbb{R}^+$ is the viscosity of the gas and $\mu \in \mathbb{R}^+$ is the permeability of the medium.

Note that Darcy's law is an empirically derived law [D] which replaces the usual conservation of momentum in the standard (Navier-Stokes) description of gas flow.

If we eliminate p and $\vec{v}$ from the equations and scale away all of the resulting constants we obtain the porous medium equation

$$\frac{\partial u}{\partial t} = \Delta(u^m) \quad , \tag{1}$$

where $m = 1 + \alpha \geq 2$. The quantity u represents a scaled density and so it is natural to assume that $u \geq 0$.

Equation (1) arises in many other applications, e.g., in the theory of ionized gases at high temperature [ZR] for values of $m > 1$, and in various models in plasma physics [BH] for values of $m < 1$. Of course, for $m = 1$ equation (1) is the classical equation of heat conduction. In these lectures I will focus on the case $m > 1$.

Other models (e.g., ground water flow) lead to equations similar to (1), but with u^m replaced by a more general nonlinear term $\varphi(u)$. I will concentrate on the porous medium case since it serves as a paradigm for the more general theory and its theory is much more complete. Finally, many problems lead to porous medium type equations with source or drift terms:

$$\frac{\partial u}{\partial t} = \Delta\varphi(u) + A\cdot\nabla\psi(u) + \sigma(u) \ .$$

Examples occur in ground water problems and in population dynamics problems. References can be found in [Ar4], [BP] and [P]. I will discuss some aspects of the population case in the last two lectures.

If we compute the Laplacian in (1) the result is

$$\Delta(u^m) = \operatorname{div}(mu^{m-1}\operatorname{grad} u) \ .$$

Thus equation (1) is uniformly parabolic in any region where u is bounded away from zero, but is degenerate in the neighborhood of any point where $u = 0$. In terms of standard Fickian diffusion theory, the diffusivity mu^{m-1} vanishes with u. The most striking manifestation of this nonlinear degeneracy is that in porous medium flow there is a finite speed of propagation of disturbances from rest. This is in stark contrast to the linear heat equation $(m = 1)$ where there is an infinite speed of propagation.

SELFSIMILAR SOLUTIONS. There are several explicit selfsimilar solutions of the porous medium equation. It is useful to look briefly at some of them since they provide a preview of much of the theory.

An important class of selfsimilar solutions can be found by assuming that u has the form

$$u(x,t) = (t_0 \pm t)^{-\alpha} f(\xi)$$

with

$$\xi \equiv x(t_0 \pm t)^{-\beta}$$

where $t_0 \in \mathbb{R}$ is arbitrary and $t_0 \pm t > 0$. The numbers α and β as well as the function f must be determined. Substituting in (1) yields

$$(t_0 \pm t)^{-m\alpha-2\beta}\Delta(f^m) = \pm(t_0 \pm t)^{-\alpha-1}(-\beta\xi\cdot\nabla f-\alpha f) .$$

Thus, if

$$(m-1)\alpha + 2\beta = 1 \tag{2}$$

we obtain a partial differential equation for $f = f(\xi)$:

$$\Delta(f^m) \pm (\beta\xi\cdot\nabla f + \alpha f) = 0 . \tag{3}$$

Some further restrictions are needed in order to fix α and β .

Barenblatt Solution [Bl]. The Barenblatt solution of (1) is a radially symmetric selfsimilar solution of the form

$$U(x,t;M) = t^{-\alpha}\{(A-B|x|^2 t^{-2\beta})_+\}^{1/(m-1)} \equiv t^{-\alpha}F(|x|t^{-\beta})$$

which satisfies

$$\int_{\mathbb{R}^d} U(x,t;M)\,dx = M \quad \text{for all} \quad t \in \mathbb{R}^+ \tag{4}$$

and arbitrary $M \in \mathbb{R}^+$. Here $(\cdot)_+ = \max(\cdot,0)$. In order that (2) and (4) hold we must have $\alpha = \beta d$ with

$$\beta = \frac{1}{2+(m-1)d} ,$$

and

$$\omega_d \int_0^\infty F(\zeta)\zeta^{d-1}d\zeta = M , \tag{5}$$

where ω_d denotes the volume of the unit ball in $\mathbb{R}^d$. If

$$B = (m-1)\beta/2m$$

then $f(\xi) = F(|\xi|)$ is a solution to (3) for $|\xi| \neq (A/B)^{1/2}$ with arbitrary $A \in \mathbb{R}^+$. Finally, the value of A is determined by the condition (5). Specifically, A must satisfy

$$\omega_d A^{1/2\beta(m-1)} B^{-d/2} \int_0^{\pi/2} (\cos\theta)^{\frac{m+1}{m-1}} (\sin\theta)^{d-1} d\theta = M .$$

The Barenblatt solution U is a classical solution of (1) and, indeed, a C^∞ function on the set

$$\mathcal{P}[U] \equiv \{(x,t) \in \mathbb{R}^d \times \mathbb{R}^+ : U(x,t) > 0\} .$$

Set

$$r(t) \equiv A^{1/2} t^\beta / B^{1/2} .$$

Then, clearly,

$$\mathcal{P}[U] = \{(x,t) \in \mathbb{R}^d \times \mathbb{R}^+ : |x| < r(t)\} .$$

The set

$$I[U] \equiv \{(x,t) \in \mathbb{R}^d \times [0,\infty) : |x| = r(t)\}$$

is called the <u>interface</u> (or free boundary) since it is the boundary of

$$\text{supp } U = \text{Cl}\mathcal{P}[U] .$$

Actually, U is a classical solution of (1) in $(\mathbb{R}^d \times \mathbb{R}^+) \setminus I$, but it is not a classical solution in all of $\mathbb{R}^d \times \mathbb{R}^+$ since $\nabla(U^{m-1})$ has jump discontinuities across I. As we shall see later on, U is a solution of (1) in $\mathbb{R}^d \times \mathbb{R}^+$ in the appropriate generalized or weak sense and is uniquely determined by the initial values on $t = 0$.

Note that $\text{supp } U(\cdot,t;M) = \text{Cl}B_{r(t)}(0)$ increases monotonically with t at the finite rate $\dot{r}(t)$ for $t > 0$. As $t \downarrow 0$, $\text{supp } U(\cdot,t)$ shrinks to the origin and it follows from (4) that $U(\cdot,t;M)dx \to M\delta_o(dx)$, i.e., initially the Barenblatt solution is a multiple of the Dirac measure concentrated at $x = 0$. It is a worthwhile exercise to show that $U(x,t;1)$ approaches the fundamental solution of the heat conduction equation as $m \downarrow 1$.

The Barenblatt solution can be embedded in a two parameter family of selfsimilar solutions by using the scale invariance properties of the porous medium equation. If $u(x,t)$ is a solution of (1), then for any positive constants p and q

$$w(x,t) \equiv \left(\frac{q}{p^2}\right)^{\frac{1}{m-1}} u(px,qt)$$

is also a solution. Set

$$W(x,t;M,p,q) \equiv \left(\frac{q}{p^2}\right)^{\frac{1}{m-1}} U(px,qt;M) .$$

Then W is a selfsimilar solution of (1) with

$$\int_{\mathbb{R}^d} W(x,t;M,p,q)dx = M\left(\frac{q}{p^{2+d(m-1)}}\right)^{\frac{1}{m-1}} ,$$

i.e., with

$$W(\cdot,t;M,p,q) \to M\left(\frac{q}{p^{2+d(m-1)}}\right)^{\frac{1}{m-1}} \delta_o(dx)$$

as $t \downarrow 0$.

It follows from the equation of state that the pressure corresponding to the scaled density u is proportional to u^{m-1}. Since pressure and velocity are related by Darcy's law, the scaled pressure

$$v \equiv \frac{m}{m-1} u^{m-1} \tag{6}$$

will play a very important role in the development of the theory. Formally, the equation for the pressure is

$$\frac{\partial v}{\partial t} = (m-1)v\,\Delta v + |\nabla v|^2 . \tag{7}$$

The pressure corresponding to the Barenblatt solution is given by

$$V(x,t;M) \equiv \frac{\beta}{2t}\{r^2(t)-|x|^2\}_+ .$$

Observe that V is continuous, while V_t and ∇V are bounded but have jump discontinuities across I.

<u>Quadratic Pressure Solution</u>. Set $\alpha = 1/(m-1)$, $\beta = 0$, and

$$t_o = \frac{m-1}{2m\{2+d(m-1)\}} .$$

We seek a selfsimilar solution of (1) of the form

$$(t_o-t)^{-\frac{1}{m-1}} f(x) = (t_o-t)^{-\frac{1}{m-1}} F(|x|) .$$

It is not difficult to verify that $f(x) = (t_o|x|^2)^{\frac{1}{m-1}}$ is a solution of (3) so that

$$\hat{U}(x,t) \equiv \left(\frac{t_o|x|^2}{t_o-t}\right)^{\frac{1}{m-1}}$$

is a solution of the required form. Note that $\hat{U}$ is a classical solution of (1) in $\mathbb{R}^d \times (0,t_o)$ with initial values

$$\hat{U}(x,0) = |x|^{\frac{2}{m-1}} .$$

However, $\hat{U}(\cdot,t) \to +\infty$ as $t \uparrow t_o$ in $\mathbb{R}^d \setminus \{0\}$. The pressure corresponding to $\hat{U}$ is given by

$$\hat{V}(x,t) \equiv \frac{mt_o|x|^2}{(m-1)(t_o-t)} .$$

Using scale invariance, we can embed $\hat{U}$ in a one parameter family of solutions. In particular, for any $q > 0$

$$\hat{W}(x,t;q) \equiv q^{\frac{1}{m-1}} \hat{U}(x,q,t) = \left(\frac{t_o|x|^2}{\frac{t_o}{q} - t}\right)^{\frac{1}{m-1}}$$

is a selfsimilar solution with quadratic pressure.

For $d = 1$, define

$$\mathcal{U}(x,t) = \begin{cases} \hat{U}(x,t) & \text{in } (-\infty,0] \times [0,t_o) \\ 0 & \text{in } \mathbb{R}^+ \times [0,t_o) . \end{cases}$$

Then, as we shall see later, $\hat{u}$ is the weak solution of (1) in $\mathbb{R}\times[0,t_o)$ with initial values

$$\hat{u}(x,o) = \begin{cases} |x|^{\frac{2}{m-1}} & \text{in } (-\infty,0] \\ 0 & \text{in } \mathbb{R}^+ . \end{cases}$$

In this case $\operatorname{supp}\hat{u} = (-\infty,0]\times[o,t_o]$ and the interface is the line segment $x = 0$ for $t\in[0,t_o]$. This example shows that the solution of (1) is not necessarily global in time, and that the support of a solution may not expand for some positive time.

Linear Pressure Solution. Take $d = 1$, $\alpha = -1/(m-1)$, $\beta = 1$ and $t_o = 0$. We look for a solution of (1) in the form

$$t^{\frac{1}{m-1}} f(\tfrac{x}{t}) \quad \text{in } \mathbb{R}\times\mathbb{R}^+ .$$

It is easy to verify that for arbitrary $\gamma > 0$

$$\tilde{U}(x,t) \equiv \{\frac{m-1}{m}\gamma t(\gamma \pm \frac{x}{t})_+\}^{\frac{1}{m-1}}$$

is such a solution. The corresponding pressure

$$\tilde{V}(x,t) = \gamma(\gamma t \pm x)_+$$

is a linear wave.

Additional information about selfsimilar solutions and further references can be found in [ZR], [B2], and [PG].

BASIC THEORY As is indicated by the various selfsimilar solutions we have discussed, we cannot expect to find a classical solution to the initial value problem for the porous medium equation, at least if we allow $u(x,0)$ to be zero at some points. Thus we need some notion of generalized solution.

To be definite, consider the 1-dimensional initial value problem

$$\begin{aligned} & u_t \equiv (u^m)_{xx} \quad \text{for } (x,t)\in\mathbb{R}\times\mathbb{R}^+ \\ & u(\cdot,0) = u_o \quad \text{for } x\in\mathbb{R}, \end{aligned} \tag{8}$$

where u_o is a given nonnegative function. A continuous nonnegative bounded function $u = u(x,t)\colon \mathbb{R}\times\mathbb{R}^+ \to [0,\infty)$ is said to be a generalized solution of (7) if $(u^m)_x$ exists and is bounded in the sense of distributions, and if for every $T\in\mathbb{R}^+$

$$\iint_{\mathbb{R}\times(0,T)} u\psi_t - \psi_x(u^m)_x + \int_{\mathbb{R}} \psi(x,0)u_o = 0$$

for all smooth test functions ψ such that $\psi = 0$ for $t = T$ and for $|x|$ large.

The basic theory for this class of generalized solutions is due to Oleinik, Kalashnikov and Czhou [OKC] (see also [O]).

Theorem 1. The generalized solution of problem (7) is unique.

Theorem 2. If u_o is a continuous nonnegative bounded function such that $(u_o^m)_x$ is bounded, then the generalized solution u of (7) exists in $\mathbb{R}\times\mathbb{R}^+$. Moreover, $u \in C^\infty(\mathcal{P})$, where

$$\mathcal{P} = \mathcal{P}[u] = \{(x,t) \in \mathbb{R}\times\mathbb{R}^+ : u(x,t) > 0\} .$$

The proof of Theorem 1 is rather standard. One writes the integral identity for the difference of two solutions u_1 and u_2, and then takes for test functions the sequence

$$\psi_n(x,t) = \alpha_n(x) \int_t^T \{(u_1)^m - (u_2)^m\} dt ,$$

where, for each n, $\alpha_n : \mathbb{R} \to [0,1]$ is a smooth function with $\alpha_n = 1$ for $|x| \le n-1$ and $\alpha_n = 0$ for $|x| \ge n$.

The proof of Theorem 2 proceeds by regularization and approximation. Let $\varphi(u) \equiv u^m$. Formally, setting $w = \varphi(u)$, problem (7) can be rewritten in the form

$$\begin{aligned} \Phi'(w)w_t &= w_{xx} \quad \text{in} \quad \mathbb{R}\times\mathbb{R}^+ \\ w(\cdot,0) &= \varphi(u_o) \quad \text{in} \quad \mathbb{R}, \end{aligned} \tag{9}$$

where $\Phi = \varphi^{-1}$. Set $z_o = \varphi(u_o)$ and construct a sequence $\{z_n\}$ of $C^\infty(\mathbb{R})$ functions with the following properties: (i) $z_n \downarrow z_o$ as $n \to \infty$ uniformly on compact intervals, and (ii) there exist positive constants K,M and m_n such that $|z_n'| \le K$ and

$$0 < m_n \le z_n \le z_{n-1} \le M$$

for all $n \ge 1$. Let $\{\gamma_n\}$ be another sequence of $C^\infty(\mathbb{R})$ functions such that $\gamma_n \in [0,1]$, $\gamma_n \equiv 1$ for $|x| \le n-2$, $\gamma_n \equiv 0$ for $|x| \ge n-1$, and $|\gamma_n'| \le K$ for all $n > 2$. For arbitrary $T \in \mathbb{R}^+$ consider the sequence of initial-boundary value problems

$$\begin{aligned} \Phi'(w)w_t &= w_{xx} \quad \text{in} \quad (-n,n) \times (0,T] \\ w(\pm n,\cdot) &= M \quad \text{in} \quad [0,T] \\ w(\cdot,0) &= \gamma_n z_n + (1-\gamma_n)M \quad \text{in} \quad [-n,n] \end{aligned} \tag{10$_n$}$$

with $n \ge 3$. These are no longer degenerate problems and so can be treated by the standard methods in the theory of quasilinear parabolic equations [LSU]. In particular, for each integer $n > 2$, problem $(10)_n$ posesses a unique classical solution $w_n(x,t)$

which satisfies $w_n \geq m_n$. The essential step in completing the proof is to show that the sequence $w_{nx} \equiv (u_n^m)_x$ is compact in the uniform topology.

An important by-product of the proof of Theorem 2 is the following result.

Comparison Principle. Let u_1 and u_2 denote solutions of problem (7) with initial values u_{10} and u_{20} respectively. Then

$$u_{10} \leq u_{20} \quad \text{in } \mathbb{R}$$

implies that

$$u_1 \leq u_2 \quad \text{in } \mathbb{R} \times \mathbb{R}^+ .$$

The Comparison Principle can be used to prove the finite speed of propagation. To do this one simply compares a solution whose initial values have bounded support with the appropriate selfsimilar solution, e.g., the Barenblatt solution or the linear pressure solution.

All of the theory which we have outlined above can be carried through with u^m replaced by a function $\varphi(u)$ satisfying $\varphi(0) = \varphi'(0) = 0$ and $\varphi'(u) > 0$ for $u > 0$ [OKC]. Moreover, Sabinina [S] has extended the results to flows in $\mathbb{R}^d$ for $d > 1$. An alternative approach based on the pressure equation (7) instead of (9) is given in [Ar2]. Benilan, Brezis and Crandall [BBC] have used nonlinear semigroup theory to construct an existence and uniqueness theory for (1) and its generalizations with $u_0 \in L^1(\mathbb{R}^d)$. Still more general existence and uniqueness theory will be described in Lecture 6.

Lecture 2

BASIC ESTIMATES. In order to study the behavior of solutions to the porous medium equation (1.1) it is necessary to have good estimates for solutions and their derivatives. In this lecture, I will describe the basic estimates which are known, starting with two which hold in any number of space dimensions and then specializing to results which hold only for $x \in \mathbb{R}$.

Maximum Principle. Let u be a solution of the initial value problem

$$\frac{\partial u}{\partial t} = \Delta(u^m) \quad \text{in } \mathbb{R}^d \times \mathbb{R}^+ \tag{1}$$

$$u(\cdot,0) = u_0 \quad \text{in } \mathbb{R}^d .$$

Then $0 \leq u_0 \leq M$ in $\mathbb{R}^d$ implies that $0 \leq u \leq M$ in $\mathbb{R}^d \times \mathbb{R}^+$.

The maximum principle is, of course, an immediate consequence of the comparison principle.

Semiconvexity of Pressure [Ar3], [AB]. Let u be a solution of problem (1) and let

$$v = \frac{m}{m-1} u^{m-1}$$

be the corresponding pressure. Then

$$\Delta v \geq -\frac{k}{t} \quad \text{in } \mathcal{D}'(\mathbb{R}^d \times \mathbb{R}^+) \tag{2}$$

where $k = (m-1+\frac{2}{d})^{-1}$. Moreover,

$$\frac{\partial v}{\partial t} \geq -\frac{(m-1)k\, v}{t} . \tag{3}$$

This result is sharp since (2) and (3) become equalities for the Barenblatt solution. It is also best possible since even for $d = 1$ it is known that, in general, v_{xx} is not bounded above [Ar2]. Circumstances in which v_{xx} is bounded above are discussed in Lecture 3.

Sketch of proof. We can assume, by approximation, that $v > 0$ and smooth. Set $p \equiv \Delta v$. Then, differentiating in the pressure equation (1.7),

$$\frac{\partial p}{\partial t} = (m-1)v\, \Delta p + 2m\, \nabla v \cdot \nabla p + (m-1)p^2 + 2 \sum_{i,j=1}^{d} \left(\frac{\partial^2 v}{\partial x_i \partial x_j}\right)^2 .$$

Observe that

$$\sum_{i,j} (a_{ij})^2 \geq \sum_i (a_{ii})^2 \geq \frac{1}{d}(\Sigma a_{ii})^2 .$$

Therefore

$$\mathcal{L}(p) \equiv \frac{\partial p}{\partial t} - (m-1)v\Delta p - 2m\nabla v \cdot \nabla p - k^{-1}p^2 \geq 0 .$$

On the other hand,

$$\mathcal{L}(-\frac{k}{t}) = \frac{k}{t^2} - k^{-1}\frac{k^2}{t^2} = 0 .$$

Thus $\mathcal{L}(-\frac{k}{t}) \leq \mathcal{L}(p)$ and we conclude that $p \geq -k/t$. The estimate for $\frac{\partial v}{\partial t}$ follows from

$$\frac{\partial v}{\partial t} = (m-1)v\, \Delta v + |\nabla v|^2 \geq (m-1)v(-\frac{k}{t}) .$$

#

Two easy, but important, consequences of the semiconvexity of v are the following.

(i) For $x \in \mathbb{R}$, (2) implies that $v_x + \frac{kx}{t}$ is a nondecreasing function of x and therefore has lateral limits everywhere in x .

(ii) It follows from (3) that $v(x_o,t_o) > 0$ implies $v(x_o,t) > 0$ for all $t > t_o$.

Velocity Estimates. Now we restrict attention to $x \in \mathbb{R}$. A by-product of the [OKC] construction of the solution of (1.8) is the boundedness of $(u^m)_x$ so that u^m is Lipschitz continuous as a function of x . The Barenblatt solution suggests that more is true, namely, that the pressure v is Lipschitz continuous in x . This is in fact true and I shall indicate two proofs, one local in time and the other global.

(A) Time local estimate:

$$|v_x(x,t)|^2 \le \frac{2}{(m+1)t}\,\|v(\cdot,t)\|_{L^\infty(\mathbb{R})} . \tag{4}$$

Proof. Fix $t \in \mathbb{R}^+$ and $y \in \mathbb{R}$. In view of (2), the function

$$\varphi(x) \equiv v(x+y,t) + \frac{x^2}{2(m+1)t}$$

is continuous, nonnegative, and convex. Therefore, for every $h \in \mathbb{R}^+$ we have

$$\varphi(x\underline{+}h) \ge \varphi(x\underline{+}h)-\varphi(x) \ge \pm\,\varphi'(x)h .$$

Assume that $\varphi'(x) \ne 0$ and choose the sign so that $\pm\,\varphi'(x)h = |\varphi'(x)|h$. Then

$$|\varphi'(x)| \le \frac{1}{h}\,\|\varphi\|_{L^\infty(x-h,x+h)} .$$

In particular, set $x = 0$ to get

$$|v_x(y,t)| \le \frac{1}{h}\Big(\|v(\cdot,t)\|_{L^\infty(\mathbb{R})} + \frac{h^2}{2(m+1)t}\Big) .$$

Now choose h to minimize the right hand side. #

(B) Let $v_o = \frac{m}{m-1}u_o^{m-1}$. Suppose v_o' is bounded in some interval $(a,b) \subset \mathbb{R}$. For any $T \in \mathbb{R}^+$ and $\delta \in (0, \frac{b-a}{2})$ let $R = (a,b)\times(0,T]$ and $R^* = (a+\delta,b-\delta)\times(0,T]$. There exists a constant $C > 0$ independent of a,b,m,δ,T and v_o such that

$$|v_x(x,t)| \le 2\|v_o'\|_{L^\infty(a,b)} + \frac{C}{\delta}\|v\|_{L^\infty(R)} . \tag{5}$$

The proof of (5) is based on Bernstein's method. Assume that v is positive and smooth. Set

$$\psi(r) = \frac{Mr}{3}(4-r) ,$$

where $M = \|v\|_{L^\infty(R)}$, and define w implicitly by the equation $v = \psi(w)$. Let $\zeta = \zeta(x)$ be a cut-off function in (a,b) with $\zeta \equiv 1$ on $(a+\delta,b-\delta)$. The proof proceeds by analysing the inequality satisfied by $\zeta^2 w_x^2$ at a maximum point. Details can be found in [Ar1].

By Darcy's law, the velocity of the gas is given by $-\frac{m}{m-1}\nabla(u^{m-1})$. Thus the estimates (4) and (5) are velocity estimates. The analogs of (4) and (5) do not hold in $\mathbb{R}^d$ for $d > 1$. In Lecture 5 I will describe a family of selfsimilar solutions which show that in $\mathbb{R}^d$, for any $d > 1$, $|\nabla v|$ can blow up in finite time. On the other hand, v is Lipschitz continuous in $\mathbb{R}^d \times (T,\infty)$ for sufficiently large $T > 0$.

<u>Estimate for</u> $|v_t|$. The estimates for $|v_x|$ given above imply that u is Holder

continuous with respect to x with exponent $\min(1,\frac{1}{m-1})$ [Arl]. Again, the Barenblatt solution shows that this result is sharp. Kruzhkov [Kr] observed that for a large class of parabolic equations, Holder continuity in x with exponent α implies Holder continuity in t with exponent $\alpha/(2+\alpha)$. Gilding [G] refined the t-exponent to $\alpha/2$. On the other hand, by assuming certain monotonicity for v_{xx}, Di Benedetto [DiB] proved that v is Lipschitz continuous in t. Both Benilan [Be], and Aronson and Caffarelli [AC2] have proved the Lipschitz continuity of v with respect to t without any assumptions on v_{xx}. I will sketch the method used in [AC2].

Theorem [AC2]. Suppose v satisfies

$$|v(x,t) - v(x',t)| \leq L_t|x-x'|$$

and

$$v_t(x,t) \geq -K_t .$$

Then there exists a constant $C>0$ depending only on m such that

$$|v_t| \leq C \max(K_t,L_t) .$$

The proof proceeds in two stages. The first is the existence of a certain selfsimilar solution of the pressure equation, and the second is a comparison argument using the selfsimilar solution.

(A) The problem

$$v_t = (m-1)vv_{xx} + v_x^2 \quad \text{in} \quad \mathbb{R}\times\mathbb{R}^+ \tag{6}$$

$$v(x,0) = |x| \quad \text{in} \quad \mathbb{R},$$

has a unique solution $\tilde{v}(x,t)$ which can be written in the form

$$\tilde{v}(x,t) = rf(\theta) \tag{7}$$

with $r = (x^2+t^2)^{1/2}$ and $\theta = \arctan x/t$, where f is even, $f'(0) = 0$, and $f''>0$.

Sketch of proof. The existence and uniqueness of the solution $v(x,t)$ of (6) is a consequence of the results of Kalashnikov [K] on solutions of the porous medium equation with unbounded data (see also Lecture 6). The representation (7) follows from a scaling argument. In particular,

$$w(x,t) \equiv \frac{1}{\lambda}\tilde{v}(\lambda x,\lambda t)$$

is also a solution of (6) for any $\lambda \in \mathbb{R}^+$. By uniqueness

$$v(x,t) = \frac{1}{\lambda}\tilde{v}(\lambda x,\lambda t)$$

for any $\lambda \in \mathbb{R}^+$ and (7) follows by setting $\lambda = 1/r$. #

(B) By approximation, we can assume that $v > 0$. For arbitrary $(x_o,t_o) \in \mathbb{R}\times\mathbb{R}^+$ with $t_o \geq 2\delta > 0$ set $\alpha \equiv v(x_o,t_o)$. The key step is to prove that $v = O(\alpha)$ in a rectangle $\{(x,t): |x-x_o| \leq A\alpha,\ 0 \leq t_o - t \leq B\alpha\}$ for some positive constants A and B which depend on δ, but are independent of x_o, t_o and α. The upper bound for v is a consequence of the upper bound for v_x and the lower bound for v_t. The lower bound for v is derived by a comparison argument based on $\tilde{v}$ and the upper bound for v_x. The function $z(x,t) \equiv \alpha^{-1} v(x_o + \alpha x, t_o + \alpha t)$ is $O(1)$ in a rectangle about $(0,0)$ whose dimensions are $O(1)$. By standard parabolic theory [LSU] there is a constant $C \in \mathbb{R}^+$ such that $|z_t(0,0)| \leq C$. The assertion of the theorem follows since $z_t(0,0) = v_t(x_o,t_o)$. #

Lecture 3

BASIC PROPERTIES OF THE INTERFACE. To fix ideas consider the initial value problem

$$\begin{aligned} u_t &= (u^m)_{xx} \quad \text{in } \mathbb{R}\times(0,T] \\ u(\cdot,0) &= u_o \quad \text{in } \mathbb{R}, \end{aligned} \tag{1}$$

for some $T \in (0,+\infty)$. For simplicity, I assume that

$$u_o \begin{cases} \equiv 0 & \text{on } \mathbb{R}^+ \\ > 0 & \text{for all sufficiently large } x < 0 . \end{cases}$$

The function

$$\zeta(t) \equiv \sup\{x \in \mathbb{R}: u(x,t) > 0\}$$

exists on $[0,T]$ with $\zeta(0) = 0$. Since $u(x_o,t_o) > 0$ implies $u(x_o,t) > 0$ for all $t \geq t_o$ it follows that $\zeta(t)$ is nondecreasing. The curve $x = \zeta(t)$ is called the (right hand) interface for u. There may be other interfaces to the left of $x = \zeta(t)$ but their basic properties are similar and I will ignore them.

(i) $\zeta(t) \in \mathrm{Lip}(0,T]$.

Proof. Fix $t_o \in (0,T)$. In Lecture 2 it was shown that there is a constant $C = C(t_o) > 0$ such that $|v_x(x,t_o)| \leq C$. By the theorem of the mean, for $x < \zeta(t_o)$ we have

$$v(x,t_o) = -\{\zeta(t_o) - x\} v_x(\bar{x},t_o) \leq C\{\zeta(t_o) - x\}$$

where $\bar{x} \in (x, \zeta(t_o))$. The assertion follows by comparing v with the linear pressure solution

$$\tilde{V}(x,t) = C\{C(t-t_o)+\zeta(t_o)-x\}_+$$

for $t \geq t_o$.

(ii) We expect that the interface will move with the local velocity of the gas. Thus, by Darcy's law, we expect

$$\dot{\zeta}(t) = -v_x(\zeta(t),t) \tag{2}$$

where

$$v_x(\zeta(t),t) \equiv \lim_{x \uparrow \zeta(t)} v_x(x,t) . \tag{3}$$

As was shown in Lecture 2, the limit in (3) exists for all t . Knerr [Kn] has shown that (2) is almost true. In particular,

$$D^+\zeta(t) = -v_x(\zeta(t),t) \tag{4}$$

holds for all $t \in (0,T]$, where D^+ denotes the right hand derivative.

<u>WAITING TIME</u>. Knerr [Kn] has shown that there exists a $t^* \in [0,T]$ such that

$$\zeta(t) \equiv 0 \quad \text{for} \quad t \in [0,\min(t^*,T)] ,$$

and

$$\zeta(t) \quad \text{is strictly increasing for} \quad t \in (\min(t^*,T),T] .$$

We call t^* the <u>waiting time</u>. It is possible to have $t^* > 0$. Indeed,

$$v_o(x) \geq C(-x)^\gamma \quad \text{on} \quad (-\delta,0) \quad \text{for some} \quad \gamma \in (0,2) \quad \text{implies} \quad t^* = 0 ,$$

while

$$v_o(x) \leq Cx^2 \quad \text{on} \quad (-\delta,0) \quad \text{implies} \quad t^* > 0 .$$

The quadratic pressure solution described in Lecture 1 shows that it is possible to have $T < \infty$ and $t^* = T$. Caffarelli and Friedman [CF1] have shown that if $t^* < T$ then

$$\zeta \in C^1(t^*,T] \quad \text{and} \quad \zeta' > 0 \quad \text{on} \quad (t^*,T] .$$

Moreover, they show that (2) actually holds for $t \in (t^*,T]$.

An important step in the proof of the smoothness of ζ given in [CF1] is a second order differential inequality satisfied by ζ which is a consequence of the semi-convexity (2.2). The sharp form of this inequality

$$\ddot{\zeta} + \frac{m}{(m+1)t}\,\dot{\zeta} \ge 0$$

was derived by Vazquez [V1]. Note that this inequality becomes an equality for the interface of the Barenblatt solution.

Estimate for t^* [ACK]. Let

$$t_m = 1/2(m+1) ,$$

and $v_0 = \frac{m}{m-1} u_0^{m-1}$. If for some $\alpha,\beta \in \mathbb{R}^+$

$$v_0(x) \le \alpha x^2 + o(x^2) \quad \text{as} \quad x \uparrow 0$$

and

$$v_0(x) \le \beta x^2 \quad \text{in} \quad \mathbb{R}^-$$

then

$$\frac{t_m}{\beta} \le t^* \le \frac{t_m}{\alpha} .$$

In particular, if $\alpha = \beta$ then $t^* = t_m/\alpha$.

Note that the upper bound for t^* is determined by "local" properties of v_0 while the lower bound is determined by "global" properties.

Example. Let $\theta \in [0,1]$ and consider the initial pressure

$$v_0(x;\theta) \equiv \begin{cases} (1-\theta)\sin^2 x + \theta \sin^4 x & \text{for } x \in [-\pi,0] \\ 0 & \text{for } x \notin (-\pi,0] . \end{cases}$$

For $\theta \in [0,1/4]$ we have $\alpha = \beta = 1-\theta$ so that $t^* = t_m/(1-\theta)$. For $\theta \in (1/4,1]$ we have $\alpha = 1-\theta < \beta$ so that we cannot evaluate t^*, but can only estimate it. In the extreme case $\theta = 1$ we have $\alpha = 0$ and

$$(1.90438\ldots)t_m \le t^* < +\infty .$$

Vazquez [V2] has derived more precise estimates for t^* in terms of integrals of the initial density. His results apply even when the initial density is given by a measure. Although I will not discuss solution of (1.1) whose initial values are measures until Lecture 6, I will describe Vazquez's results on the waiting time here.

Consider the initial value problem (1), where now u_0 is a nonnegative measure on $\mathbb{R}$. Assume that

$$\sup\{x \in \mathbb{R} : \int_x^\infty u_o(dx) > 0\} = 0 ,$$

i.e., the gas is initially concentrated on $\mathbb{R}^-$ and every left neighborhood of $x = 0$ contains some gas. The initial values are taken on in the sense of trace:

$$\lim_{t \downarrow 0} \int_{\mathbb{R}} \psi(x)u(x,t)\,dx = \int_{\mathbb{R}} \psi(x)u_o(dx)$$

for all test functions $\psi \in C_o(\mathbb{R})$. This initial value problem has a unique continuous weak solution in $\mathbb{R} \times (0,T)$ for some $T > 0$ if and only if

$$\ell \equiv \limsup_{r \to \infty} r^{-\frac{m+1}{m-1}} \int_{|y| \le r} u_o(dy) < \infty .$$

Roughly speaking, $\ell < \infty$ means that, on average, the initial pressure does not grow faster than x^2. The maximal time of existence, T, is determined by ℓ with $T = \infty$ if and only if $\ell = 0$. For more details see Lecture 6.

Set

$$M(x) \equiv \int_{(x,0]} u_o(d\xi)$$

and define

$$B \equiv \sup_{x<0} |x|^{-\frac{m+1}{m-1}} M(x) , \quad A_+ \equiv \limsup_{x \uparrow 0} |x|^{-\frac{m+1}{m-1}} M(x) ,$$

and

$$A_- \equiv \liminf_{x \uparrow 0} |x|^{-\frac{m+1}{m-1}} M(x) .$$

Clearly $0 < B \le 0$, $0 \le A^- \le A^+ \le B$, and $B = \infty$ if and only if $A^+ = \infty$. In addition, define

$$T_m \equiv \frac{1}{2m}\left(\frac{m-1}{m+1}\right)^m , \quad T_B \equiv T_m B^{1-m} , \text{ and } T_{A_\pm} \equiv T_m A_\pm^{1-m} .$$

Note that $T_{A_\pm}$ depends on local properties of u_o while T_B depends on global behavior. In the case considered by [ACK] and described above $A_+ = A_- = A$, and $T_A = 1/2(m+1)\alpha$, $T_B = 1/2(m+1)\beta$.

Theorem [V2]. (i) There exists a constant $\mu = \mu(m) > 1$ such that

$$T_B \le t^* \le \mu T_B$$

(ii) If $A_- > 0$ then $t^* \le T_{A_-}$.

Note that, according to (i), $t^* > 0$ if and only if $T_B > 0$, i.e., if and only if $B < \infty$. Moreover, if $B = A_-$ then $t^* = T_B$.

REGULARITY AND IRREGULARITY. Suppose that $t^* \in (0,T)$. Then $\zeta \in C^1$ for $t \in (0,t^*) \cup (t^*,T)$. We now ask if ζ is a C^1 function in a neighborhood of $t = t^*$? This will be the case if $D^-\zeta(t^*) = D^+\zeta(t^*)$, i.e., if $D^+\zeta(t^*) = 0$. If $D^+\zeta(t^*) \neq 0$, then the interface has a corner point at $t = t^*$. As the next result shows, both cases can occur.

Theorem [ACV]. (i) If $t^* = T_B$ then $\zeta \in C^1(0,T)$.

(ii) If $t^* < T$, $A_+ < \infty$, and $t^* < T_{A_+}$ then $\zeta \notin C^1(0,T)$.

If $A_+ = A_- = B$ then $t^* = T_B$ and $\zeta \in C^1(0,T)$. On the other hand, if $t^* = T_B < T$ then, according to (i), $\zeta \in C^1(0,T)$ and it follows from (ii) that $t^* \geq T_{A_+}$. However, by definition, $T_{A_+} \geq T_B$. Therefore $t^* = T_B < T$ implies that $A_+ = B$.

If $t^* < T$ and $M(x) = o(|x|^{\frac{m+1}{m-1}})$ as $x \uparrow 0$ then $\zeta \notin C^1(0,T)$. In particular, this will be the case if $v_0(x) = o(x^2)$ as $x \uparrow 0$.

Example. In the "sine" example discussed above, if $\theta \in [0,1/4]$ then $A_+ = A_- = B$ and $\zeta \in C^1(\mathbb{R})$. If $\theta = 1$, then $v_0 = o(x^2)$ as $x \uparrow 0$ and there is a corner point at $t = t^*$. This is shown very strikingly by the computations of Tomoeda and Mimura [TM] given in Figure 3.1. I conjecture that $\zeta \notin C^1(\mathbb{R})$ for all $\theta \in (1/4,1]$.

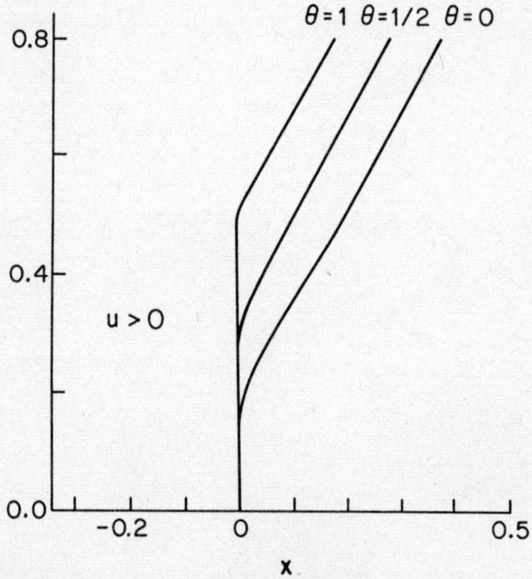

Figure 3.1. The interface for the "sine" example for $\theta = 0, 1/2$ and 1 as computed by Tomoeda and Mimura [TM].

In the "sine" example for $\theta=0$, i.e., for

$$v_o(x) = \begin{cases} \sin^2 x & \text{on } (-\pi,0) \\ 0 & \text{on } \mathbb{R} \quad (-\pi,0) , \end{cases}$$

one can calculate $v_{xx}(0,t)$ explicitly [Ar2]. In fact

$$v_{xx}(0,t) = \frac{2t^*}{t^* - t}$$

so that $v_{xx}(0,t) \to \infty$ as $t \uparrow t^*$ even though the interface belongs to $C^1(\mathbb{R}^+)$. On the other hand, the following result holds when $\zeta \notin C^1(0,T)$.

Theorem [ACV]. Suppose that $t^* \in (0,T)$, $D^+\zeta(t^*) = \gamma > 0$, and $u>0$ for (x,t) near $(0,t^*)$. Then for every $\theta \in (0,\gamma)$

$$\lim_{\epsilon \downarrow 0} \{ \sup_{|x+\gamma\epsilon| < \theta\epsilon} \epsilon v_{xx}(x,t^*-\epsilon)\} = +\infty .$$

These results suggest that v_{xx} is always unbounded near $(\zeta(t^*),t^*)$. It is therefore not too surprising that v_{xx} is locally bounded away from $(\zeta(t^*),t^*)$.

Theorem [AV2]. If $t^* \in (0,T)$ then v_{xx} is bounded in a neighborhood in supp u of every point $(\zeta(t_o),t_o)$ with $t_o \in (t^*,T)$.

The proof of this result is based on a barrier argument for v_{xx} with a barrier of the form

$$p(x,t) = \frac{\alpha}{\zeta(t)-x} + \frac{\beta}{\zeta^*(t)-x}$$

where α and β are positive constants, and $\zeta^* > \zeta$ in some neighborhood of t_o . The fact that $\dot{\zeta}(t_o) = -v_x(\zeta(t_o),t_o) \equiv a > 0$ allows one to show that for α and β sufficiently small, p is a supersolution of the differential equation for v_{xx} in a suitable neighborhood of $(\zeta(t_o),t_o)$. On the other hand, Caffarelli and Friedman [CF1] have shown that $vv_{xx} \to 0$ at points on the moving interface. Using this, one can show that $p \geq v_{xx}$ on the parabolic boundary of a suitable neighborhood of $(\zeta(t_o),t_o)$ for arbitrarily small $\alpha > 0$. An upper bound for v_{xx} is obtained by letting $\alpha \downarrow 0$. The lower bound for v_{xx} comes from the semiconvexity property (Lecture 2).

Using the boundedness of v_{xx} and a somewhat more elaborate barrier argument, one can bound all of the derivatives of v near the moving boundary.

Theorem [AV2]. Every point $(\zeta(t),t)$ on the interface with $t\in(t^*,T)$ has a neighborhood η such that v is a C^∞ function of (x,t) in $\eta\cap(\text{supp } u)$ and ζ is a C^∞ function of t in $\eta \cap \{(x,t): x=\gamma(t)\ ,\ t\in(t^*,T)\}$.

In the case that v_o has compact support, the fact that $\zeta\in C^\infty(t^*,\infty)$ was first proved by Hollig and Kreiss [HK] using some weighted integral estimates. In this case there is also another proof due to Angenent [An] using nonlinear semigroup theory which shows that ζ and the corresponding left hand interface are analytic functions on (t^*,∞) .

ASYMPTOTIC BEHAVIOR. It is possible to give a very precise description of the asymptotic form of the interface if $\text{supp } u_o = (a_1,a_2)$ with $-\infty<a_1<a_2<+\infty$. We now consider both the left and right hand interfaces $x = \zeta_i(0) = a_i$. Let

$$M = \int_{a_1}^{a_2} u_o \quad \text{and} \quad x_o = \frac{1}{a_2-a_1}\int_{a_1}^{a_2} x\, u_o \ .$$

Theorem [V1]. As $t\to\infty$

$$\zeta_i(t) = x_o + (-1)^i A^{1/2}B^{-1/2}t^{\frac{1}{m+1}} + o(1)$$

and

$$\dot{\zeta}_i(t) = \frac{(-1)^i}{m+1} A^{1/2}B^{-1/2}t^{-\frac{m}{m+1}} + o(1/t)\ ,$$

where A and B are the constants in the Barenblatt solution. Specifically,

$$B = \frac{m-1}{2m(m+1)}$$

and A is determined by M .

The proof of this result involves very delicate comparisons with the Barenblatt solution of mass M centered at x_o .

Lecture 4

CONNECTIONS WITH HAMILTON-JACOBI EQUATIONS. If $u=u(x,t)$ is the solution of the problem

$$u_t = (u^m)_{xx} \quad \text{in} \quad Q \equiv \mathbb{R}\times \mathbb{R}^+$$
$$u(\cdot,0) = u_o \quad \text{in} \quad \mathbb{R}\ ,$$

where $u_o\in L^1(\mathbb{R})$, then, according to a theorem of Benilan and Crandall [BC], u depends continuously in the $C(\mathbb{R}^+ : L^1(\mathbb{R}))$-norm on both u_o and m . In particular, if u_o is held fixed and $m\downarrow 1$ then u converges to the solution of

$$u_t = u_{xx} \quad \text{in } Q$$

$$u(\cdot,0) = u_0 \quad \text{in } \mathbb{R} .$$

Thus for m near 1 the porous medium equation can be regarded as a perturbation of the heat conduction equation.

Despite the convergence of solutions of the nonlinear porous medium equation to solution of the linear heat conduction equation, there is a marked difference in the behavior of solutions to these equations. In particular, as we have seen in detail in Lecture 3, there is a finite speed of propagation associated with the porous medium equation. This is, of course, reminiscent of the behavior of solutions of hyperbolic equations. To explore this connection further we introduce the pressure $v = mu^{m-1}/(m-1)$ which satisfies the equation

$$v_t = (m-1)vv_{xx} + v_x^2 . \tag{1}$$

Formally, as $m \downarrow 1$, v satisfies the Hamilton-Jacobi equation

$$v_t = (v_x)^2 . \tag{2}$$

In this lecture, I will show how to make these formal observations precise.

Set $\epsilon = m-1$. For each $\epsilon > 0$ let $v_{\epsilon 0}$ be a continuous nonnegative function and consider the initial value problem

$$v_t = \epsilon vv_{xx} + (v_x)^2 \quad \text{in } Q \tag{3$_\epsilon$}$$

$$v(\cdot,0) = v_{\epsilon 0} \quad \text{in } \mathbb{R} .$$

A solution to problem $(3)_\epsilon$ is the pressure computed from the generalized solution to the initial value problem for equation (1.1) with $u_0 = \{(m-1)v_{\epsilon 0}/m\}^{1/(m-1)}$. For $\tau \in \mathbb{R}$ we write $Q_\tau \equiv \{(x,t) \in Q\colon t > \tau\}$.

Theorem 1 [AV1]. (A) Suppose that for all sufficiently small $\epsilon > 0$ there is a constant $N > 0$ independent of ϵ such that

$$0 \le v_{0\epsilon} \le N$$

and suppose that $v_{\epsilon 0} \to v_0$ as $\epsilon \downarrow 0$ uniformly on compact subsets of $\mathbb{R}$. Let v_ϵ denote the solution of problem $(3)_\epsilon$. Then as $\epsilon \downarrow 0$ the family $\{v_\epsilon\}$ converges uniformly on compact subsets of $\bar{Q}$ to a function $v \in C(\bar{Q})$ such that

(i) $v \in \mathrm{Lip}(Q_\tau)$ for every $\tau > 0$ and $v_t = (v_x)^2$ a.e. in Q .

(ii) $v(x,0) = v_0(x)$ for all $x \in \mathbb{R}$.

(iii) $v_{xx} \geq -1/2t$ in $\mathcal{D}'(Q)$.

Moreover, $v_{\epsilon x} \to v_x$ in $L^p_{loc}(Q)$ for every $p \in [1,\infty)$.

(B) The limit function $v = v(x,t)$ is uniquely characterized as a solution in $C(\bar{Q})$ of the initial value problem

$$\begin{aligned} &v_t = (v_x)^2 \quad \text{in } Q \\ &v(\cdot,0) = v_o \quad \text{in } \mathbb{R} \end{aligned} \tag{4}$$

by the semiconvexity property (iii).

I sketch the proof of the existence of a limit as $\epsilon \downarrow 0$. By the maximum principle, the family $\{v_\epsilon\}$ is uniformly bounded in Q . Moreover, by the estimates given in Lecture 2, the family $\{v_\epsilon\}$ is equicontinuous in Q_τ for any $\tau > 0$. (Actually one must check that the Lipschitz constants for v with respect to x and t do not blow up as $\epsilon \downarrow 0$. This is done in [AV1].) Thus there exists a sequence $\epsilon_n \downarrow 0$ such that $v_n \equiv v_{\epsilon_n} \to v \in C(Q)$ uniformly on Q_τ for every $\tau > 0$. The limit function v satisfies

$$\begin{aligned} &0 \leq v \leq \|v_o\|_{L^\infty(\mathbb{R})} \quad \text{in } Q , \\ &|v_x|^2 \leq \frac{2\|v_o\|_{L^\infty(\mathbb{R})}}{t} \quad \text{in } Q , \\ &v_{xx} \geq -1/2t \quad \text{in } \mathcal{D}'(Q) , \end{aligned} \tag{5}$$

and

$$v_t \geq 0 \quad \text{in } \mathcal{D}'(Q) .$$

Moreover, it can be shown that v is uniformly Lipschitz continuous with respect to (x,t) in Q_τ for any $\tau > 0$. By a slight variant of a lemma due to P. L. Lions [L], (5) implies that $\{v_{nx}\}$ is relatively compact in $L^p_{loc}(Q)$ for every $p \in [1,\infty)$. Thus (on a suitable subsequence)

$$v_{nx} \to v_x \quad \text{in } L^p_{loc}(Q) \text{ and a.e.}$$

Since the v_n satisfy

$$\int_Q \{\epsilon_n v_n v_{nx} \psi_x + (\epsilon_n - 1)(v_{nx})^2 \psi - v_n \psi_t\} = 0$$

for all test functions $\psi \in C^1_o(Q)$, we obtain in the limit

$$\int_Q (v_x)^2 \psi + v\psi_t = 0 .$$

Thus v satisfies (2) in $\mathcal{D}'(Q)$ and a.e. #

The local velocity of propagation for solutions to the porous medium equation is given by $w = -v_x$. Theorem 1 implies that the family $\{w_\varepsilon\} = \{-v_{\varepsilon x}\}$ converges in $L^p_{loc}(Q)$ to a solution w of the <u>conservation law</u>

$$w_t + (w^2)_x = 0 \quad \text{in} \quad \mathcal{D}'(Q)$$

and that w satisfies the <u>entropy condition</u>

$$w_x \le 1/2t \ .$$

If, in addition, v_{ox} exists in a suitable sense, then w is the unique distribution solution with these properties.

Crandall and P. L. Lions [CL] have recently introduced the notion of <u>viscosity solution</u> to characterize the "good" solutions of Hamilton-Jacobi equations. Consider the equation

$$u_t + H(\nabla u) = 0 \quad \text{in} \quad Q_T \equiv \mathbb{R}^d \times (0,T) \tag{6}$$

with $H: \mathbb{R}^d \to \mathbb{R}$. A function $u \in C(Q_T)$ is said to be a <u>viscosity solution</u> of (6) if for every $\varphi \in C^1(Q_T)$ we have

$$\varphi_t + H(\nabla\varphi) \le 0$$

at all local maxima of $u-\varphi$ and

$$\varphi_t + H(\nabla\varphi) \ge 0$$

at all local minima of $u-\varphi$. By a Theorem of P. L. Lions [L] the local boundedness of v_x and semiconvexity (iii) imply that our limit function v is a viscosity solution of (2).

Consider problems $(3)_\varepsilon$ with fixed initial data $v_{\varepsilon o} \equiv v_o$, where $v_o = 0$ in $\mathbb{R}^+$ and $0 = \sup\{x \in \mathbb{R} : v_o(x) > 0\}$. Let

$$\zeta_\varepsilon(t) \equiv \sup\{x \in \mathbb{R} : v_\varepsilon(x,t) > 0\}$$

and

$$\zeta(t) \equiv \sup\{x \in \mathbb{R} : v(x,t) > 0\} \ ,$$

i.e., $x = \zeta_\varepsilon(t)$ and $x = \zeta(t)$ are the (right hand) interfaces for problems $(3)_\varepsilon$ and (4) respectively.

Theorem 2 [AV1]. (A) As $\epsilon \downarrow 0$ the family $\{\zeta_\epsilon\}$ converges uniformly on compact subsets of $[0,\infty)$ to ζ.

(B) $\dot{\zeta}_\epsilon \to \dot{\zeta}$ in $L^p_{loc}(\mathbb{R}^+)$ for every $p \in [1,\infty)$ and ζ satisfies

$$\ddot{\xi} + \frac{1}{2t}\dot{\xi} \geq 0 \quad \text{in} \quad \mathcal{D}'(\mathbb{R}^+) .$$

(C) $t^*_\epsilon \to 1/4D$ where $D = \sup_{x<0} \{|x|^{-2} v_0(x)\}$.

(D) For all $t > 0$ and every sequence $\epsilon_n \downarrow 0$ such that $\dot{\zeta}_{\epsilon_n}(t)$ converges we have

$$\lim \dot{\zeta}_{\epsilon_n}(t) \in [\dot{\zeta}(t-),\dot{\zeta}(t+)] .$$

In particular, $\dot{\zeta}_\epsilon \to \dot{\zeta}$ at every point where $\dot{\zeta}$ exists.

Roughly speaking, Theorems 1 and 2 say that problems $(3)_\epsilon$ give finite speed approximations to the Hamilton-Jacobi problem (4) with convergence of the interfaces. In contrast, the usual artificial viscosity methods yield approximations with infinite propagation speed. This observation might be useful in finidng efficient numerical methods for dealing with problems such as (4).

Finally, I want to discuss another connection between equations (1) and (2). By the semiconvexity of the pressure we have

$$v_t + \frac{m-1}{(m+1)t} v \geq (v_x)^2 \quad . \tag{7}$$

Define

$$Z(x,\tau) \equiv v(x,t)t^{\frac{m-1}{m+1}} , \tag{8}$$

where

$$\tau = \frac{m+1}{2} t^{\frac{2}{m+1}} \quad . \tag{9}$$

With this change of variables, (7) and the semiconvexity (2.2) become

$$Z_\tau \geq (Z_x)^2 \quad \text{and} \quad Z_{xx} \geq -\frac{1}{2\tau} . \tag{10}$$

Thus, if u is a solution of the porous medium equation (1.1) then the change of variables (8),(9) transforms the corresponding pressure into a viscosity supersolution of equation (2).

The equalities in (7) and (10) hold for the Barenblatt solution $U(x,t;M)$ introduced in Lecture 1. With the change of variables (8),(9) the Barenblatt pressure V becomes

$$Z(x,y) = \left(\frac{mA}{4(m-1)} - \frac{x^2}{4\tau}\right)_+ \tag{11}$$

The functions (11) are known bounded selfsimilar solutions to

$$Z_\tau = (Z_x)^2$$

with initial values

$$Z(x,0) = \begin{cases} 0 & \text{if } x \neq 0 \\ \frac{mA}{4(m-1)} & \text{if } x = 0 . \end{cases}$$

This correspondence has some interesting consequences. It is known that as $t \to \infty$ every solution u of the porous medium equation with $u_0 \in L^1(\mathbb{R})$, $u_0 \geq 0$, and $u_0 \neq 0$ converges with the appropriate scaling to the Barenblatt solution U with the same mass i.e., $\int u_0 = \int U = M$. Specifically, Kamin [Ka] has shown that

$$t^{1/m+1} \|u(\cdot,t) - U(\cdot,t;M)\|_{L^\infty(\mathbb{R})} \quad \text{as} \quad t \to \infty .$$

The selfsimilar solutions (11) play the same role for bounded solutions of (2). Thus the asymptotic behavior of these classes of solutions to the porous medium equation and the Hamilton-Jacobi equation (2) coincide under the transformation $v = \frac{m}{m-1} u^{m-1}$ together with (8) and (9).

Lions, Souganidis and Vazquez [LSV] have extended parts of Theorems 1 and 2 to the case $x \in \mathbb{R}^d$ for $d > 1$.

Lecture 5

HOLDER CONTINUITY IN $\mathbb{R}^d$. Caffarelli and Friedman [CF2], [CF3] have studied the regularity of solutions of the porous medium equation and of the interface for flows in $\mathbb{R}^d$ with $d > 1$. Their results are certainly not optimal and an important open problem is to find the optimal regularity results in $\mathbb{R}^d$.

Consider the initial value problem

$$\begin{aligned} \frac{\partial u}{\partial t} &= \Delta(u^m) \quad \text{in} \quad \mathbb{R}^d \times \mathbb{R}^+ \\ u(\cdot,0) &= u_0 \quad \text{in} \quad \mathbb{R}^d \end{aligned} \tag{1}$$

where $u_0 : \mathbb{R}^d \to [0,N]$ and $u_0 \in L^2(\mathbb{R}^d)$. Sabinina [S] has proved the existence of a unique generalized solution u of this problem with u, $\nabla u^m \in L^2(\mathbb{R}^d \times (0,T))$ for any $T > 0$.

The tools Caffarelli and Friedman use are the semiconvexity estimates (2.2), (2.3) and two lemmas about averages which are derived from semiconvexity and scaling. Roughly speaking, these lemmas are the following:

(i) If there is no gas in the ball $B_R(x^o)$ at time $t = t^o$ and if the total amount of gas in $B_R(x^o)$ is sufficiently small at time $t = t^o+\sigma$, then there is no gas in $B_{R/6}(x^o)$ at time $t = t^o+\sigma$.

(ii) If there is a sufficiently large mass in $B_R(x^o)$ at time $t = t^o$ then the gas must cover a neighborhood of x^o at time $t = t^o+\sigma$.

Theorem 1 [CF3]. The solution u of problem (1) is uniformly Holder continuous in $\mathbb{R}^d \times (\delta,\infty)$ for any $\delta > 0$.

The proof of Theorem 1 does not give any useful estimate for the Holder exponents.

Suppose that u_o has compact support in $\mathbb{R}^d$. Define

$$\Omega(t) \equiv \{x \in \mathbb{R}^d : u(x,t) > 0\}$$

and the interface

$$\Gamma(t) \equiv \partial\Omega(t) .$$

The lower bound (2.3) for v_t implies that $\Omega(t) \subset \Omega(t')$ for $t \leq t'$, but we know that $\Omega(t)$ is not necessarily strictly increasing. Indeed, Chipot and Sideris [CS] have derived local lower bounds for the waiting time. For arbitrary $(x^*,t^*) \in \mathbb{R}^d \times \mathbb{R}^+$ define

$$\sigma(x^*,t^*) \equiv \{(x,t) \in \mathbb{R}^d \times \mathbb{R}^+ : x = x^*, 0 < t < t^*\} .$$

Theorem 2 [CF3]. For any point $(x^*,t^*) \in \Gamma(t^*)$ either

$$\sigma(x^*,t^*) \subset \Gamma \quad \text{or} \quad \sigma(x^*,t^*) \cap \Gamma = \emptyset$$

where $\Gamma \equiv \bigcup_{\mathbb{R}^+} \Gamma(t)$.

The Holder growth of the interface is given by:

Theorem 3 [CF3]. Suppose $(x^*,t^*) \in \Gamma(t^*)$ and $\sigma(x^*,t^*) \cap \Gamma = \emptyset$. Then there exist positive constants C, γ, and h independent of (x^*,t^*) such that

$$u(x,t) = 0 \quad \text{if} \quad |x-x^*| \leq C(t^*-t)^\gamma \quad \text{and} \quad t \in (t^*-h,t^*)$$

and

$$u(x,t) > 0 \quad \text{if} \quad |x-x^*| \leq C(t-t^*)^\gamma \quad \text{and} \quad t \in (t^*,t^*+h) .$$

Now assume that supp u_o is smoothly bounded and that

$$u_o(x) \geq C_o\{d(x, \partial \operatorname{supp} u_o)\}^\delta$$

for some constants $C_o \in \mathbb{R}^+$ and $\delta \in (0,2)$. Just as in the case $d = 1$ it can be shown that $\operatorname{supp} u(\cdot,t)$ is strictly increasing for $t \geq 0$. By Theorem 2, Γ is given by a function $t = S(x)$. Since $\Gamma(t)$ is monotone, contains no vertical segements and is closed, it follows that S is continuous. By comparison with a Barenblatt solution, one can show that $\inf\{|x|: x \in \Gamma(t)\} \to \infty$ as $t \to \infty$. Hence S is defined throughout $\mathbb{R}^d$. Finally, Theorem 3 implies that

$$|S(x)-S(x')| \leq C|x-x'|^{1/\gamma} .$$

ULTIMATE OPTIMAL REGULARITY. For $d = 1$ we have seen in Lecture 2 that the pressure v is always Lipschitz continuous for $t > 0$. Recently Caffarelli, Vazquez and Wolanski [CVW] have shown that the same is true in $\mathbb{R}^d$ provided that t is sufficiently large. Specifically, they prove the following. Suppose that

$$\operatorname{supp} u_o \subset B_r(0) .$$

Although $\operatorname{supp} u(\cdot,t)$ is bounded for each $t \in \mathbb{R}^+$ it does spread out and eventually covers all of $\mathbb{R}^d$. Let T be such that

$$B_r(0) \subset\subset \operatorname{supp} u(\cdot,T) .$$

Then it is proved in [CVW] that ∇v and v_t are bounded in $\mathbb{R}^d \times (\tau,\infty)$ for every $\tau > T$. The bounds depend only on τ and v_o. A consequence of this result is that for $t > T$ the interface Γ can be represented in the form $r = f(\theta,t)$, where (r,θ) are spherical coordinates in $\mathbb{R}^d$. Caffarelli et al. prove that f is a Lipschitz continuous function: $S^{d-1} \times (T,\infty) \to \mathbb{R}$.

Caffarelli and Wolanski [CW] have extended these results to show that $v \in C^{1+\alpha}$ on $\operatorname{supp} u$ for $t > T$.

A COUNTEREXAMPLE. The Caffarelli-Vazquez-Wolanski result described above is, in some sense, the best possible as is shown by the following example which was found heuristically and numerically by J. Graveleau in 1973 [Gr]. Graveleau's example shows that if there are holes in $\operatorname{supp} u_o$ then it is possible for ∇v to blow up. Thus v cannot, in general, be Lipschitz continuous in $\mathbb{R}^d \times (\tau,\infty)$ for arbitrarily small $\tau > 0$. The existence and uniqueness of Graveleau's solution is proved in [AG] and I will briefly describe it here.

We consider a radially symmetric porous medium flow in $\mathbb{R}^d$ and suppose that initially the gas lies completely outside some ball about the origin. As time increases the gas will flow into the ball and ultimately reach the center. We are interested in the behavior of the pressure at $r = 0$ at the instant the gas reaches there. Graveleau's example shows that there exist solutions such that at the instant of focusing $v \sim r^p$ where $p = p(d,m) \in (0,1]$. For $d = 1$, the Graveleau solution consists of a pair of colliding linear pressure solutions so that $p(1,m) = 1$. For $d > 1$ the value of p can only be found numerically. For example,

$$P(2,2) = 0.832221204\ldots .$$

For simplicity, we consider only the case $m = 2$, i.e.,

$$\frac{\partial u}{\partial t} = \Delta(u^2) .$$

For general $m > 1$ one deals with the pressure v rather than u. We seek a radially symmetric solution $u = u(|x|,t)$ in $\mathbb{R}^d \times \mathbb{R}^-$ with $u(|x|,0) = |x|^{2-\alpha}$ for some $\alpha \in [1,2)$. Assume that

$$u(|x|,t) = -\frac{r^2}{t}\varphi(\eta) \tag{3}$$

where

$$\eta = |x|^{-\alpha}t .$$

Then

$$-\frac{r^2}{t}\varphi(\eta) = -r^{2-\alpha}\frac{\varphi(\eta)}{\eta}$$

and we require that

$$\varphi(o) = 0 \quad \text{and} \quad \varphi'(o) = -1 . \tag{4}$$

The problem is to find the constant α and the function φ.

If we substitute (3) in (2) we obtain the ordinary differential equation

$$2\alpha^2(\varphi\varphi'' + \varphi'^2) - 2\alpha(6-\alpha+d)\frac{\varphi\varphi'}{\eta} + k\frac{\varphi^2}{\eta^2} = \frac{\varphi}{\eta^2} - \frac{\varphi'}{\eta} , \tag{5}$$

where $' = d/d\eta$ and $k = 4(d+2)$. Set

$$\eta = -e^{-\xi}$$

and $\varphi(\eta) = \varphi(-e^{-\xi}) \equiv \psi(\xi)$. Then (5) becomes

$$\psi\psi'' + \psi'^2 + \frac{6+d}{\alpha}\psi\psi' + \frac{k}{2\alpha^2}\psi^2 = \frac{1}{2\alpha^2}(\psi+\psi') , \tag{6}$$

where now $' = d/d\xi$. Moreover, the conditions (4) become

$$\psi(\xi) \to 0 \quad \text{and} \quad \psi'(\xi) \sim -e^{-\xi} \quad \text{as} \quad \xi \to \infty . \tag{7}$$

Equation (6) is singular at $\psi = 0$ so we resort to a nonlinear change of independent variable to eliminate the singularity. Write (6) as a first order system

$$\psi' = \theta$$

$$\psi\theta' = \frac{1}{2\alpha^2}\psi + \frac{1}{2\alpha^2}\theta - \frac{k}{2\alpha^2}\psi^2 - \frac{6+d}{\alpha}\psi\theta - \theta^2$$

and introduce a new variable τ defined by

$$\frac{d\tau}{d\xi} = \frac{1}{\psi(\xi)} .$$

Then with $\cdot = \frac{d}{d\tau}$ the system becomes

$$\dot\psi = \psi\theta$$

$$\dot\theta = \frac{1}{2\alpha^2}\psi + \frac{1}{2\alpha^2}\theta - \frac{k}{2\alpha^2}\psi^2 - \frac{6+d}{\alpha}\psi\theta - \theta^2 . \tag{8}$$

The system (8) has rest points at

$$(\psi,\theta) = (1/k,0) , (0,0) , \text{ and } (0,1/2\alpha^2) ,$$

where $(0,0)$ and $(0,1/2\alpha^2)$ are saddle points. In view of (7), the solution we seek corresponds to the part of the stable manifold of $(0,0)$ which lies in the half-space $\psi > 0$. For α small, as $\tau \downarrow -\infty$ this manifold either approaches the rest point $(0,0)$ (Figure 5.1a) or an unstable limit cycle which results from a Hopf bifurcation at $(0,0)$ (Figure 5.1b). For α large, as $\tau \downarrow -\infty$ the stable manifold of $(0,0)$ is asymptotic to the positive θ axis (Figure 5.1d). For an intermediate value of $\alpha, \alpha = \alpha^*$, the stable manifold of $(0,0)$ coincides with the unstable manifold of the saddle point $(0,1/2\alpha^{*2})$, i.e., there is a heteroclinic connection between the two saddles (Figure 5.1c). Translated back into the original variables, this heteroclinic connection gives the Graveleau solution. Because of the singular nature of the problem, the conditions at $(0,1/2\alpha^{*2})$ correspond to the existence of an $A = A(d) \in \mathbb{R}^+$ such that $\varphi(-A) = 0$, i.e., there exists an interface given by $|x|^{-\alpha^*} t = -A$ (Figure 5.2). For details consult [AG]

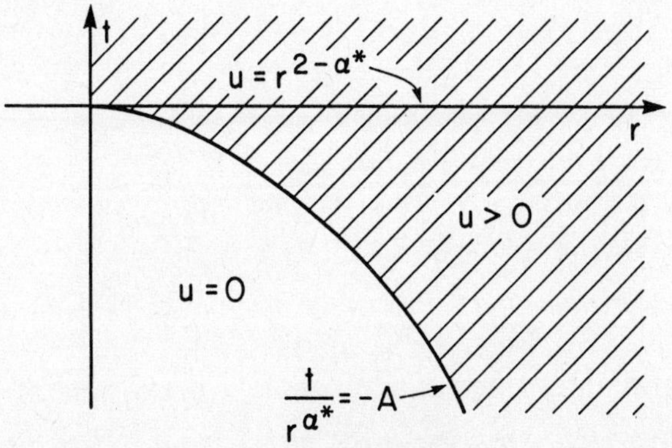

Figure 5.2. The shaded region is the support of the Graveleau solution.

I conjecture that the Graveleau solution gives the optimal regularity in x for v . This problem is under intense study.

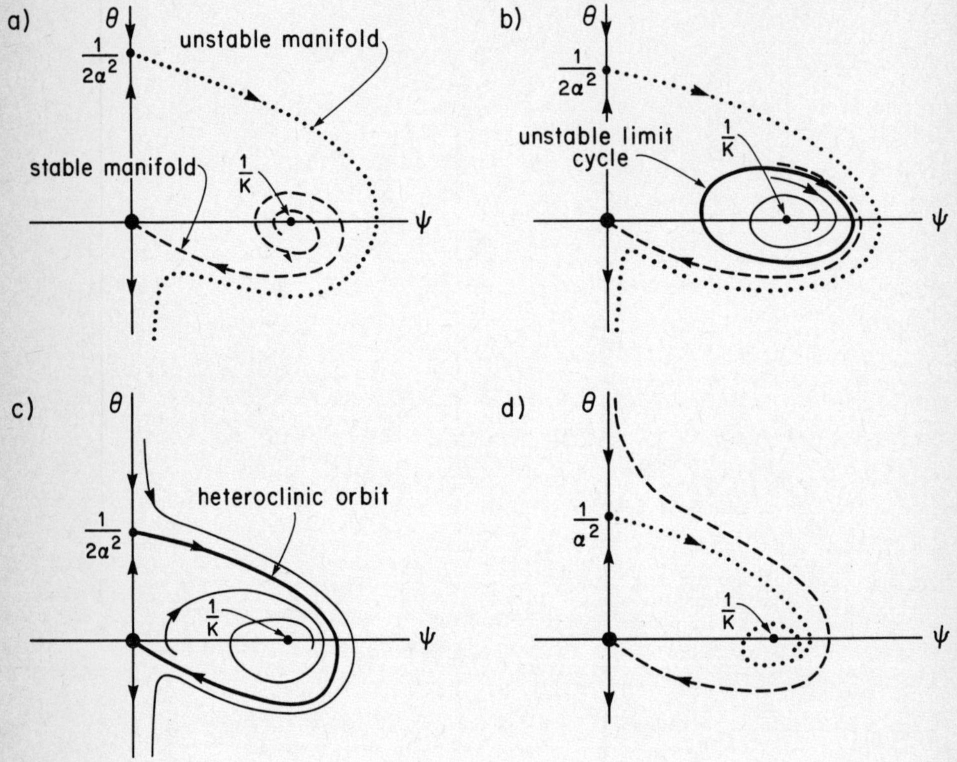

Figure 5.1. The phase plane for equation (8). (a) $\alpha \in (0,\alpha^*)$, before the Hopf bifurcation at $(\frac{1}{k},0)$. (b) $\alpha \in (0,\alpha^*)$ after the Hopf bifurcation at $(\frac{1}{k},0)$. (c) $\alpha = \alpha^*$ with the saddle-saddle connection. (d) $\alpha > \alpha^*$.

Lecture 6

INITIAL TRACE THEORY. Let $u = u(x,t)$ be a nonnegative solution of the equation of heat conduction

$$\frac{\partial u}{\partial t} = \Delta u \quad \text{in} \quad S_T \equiv \mathbb{R}^d \times (0,T] \tag{1}$$

for some $T > 0$. A consequence of Widder's representation theorem [W] is the existence of a unique nonnegative Borel measure ρ such that

$$\lim_{t \downarrow 0} \int_{\mathbb{R}^d} u(x,t)\psi(x)dx = \int_{\mathbb{R}^d} \psi(x)\rho(dx) \tag{2}$$

for all test functions $\psi \in C_0(\mathbb{R}^d)$. That is, every nonnegative solution u of (1) has a unique Borel measure ρ as <u>initial trace</u> in the sense of (2). The measure ρ is σ-finite and satisfies the growth condition

$$\int_{\mathbb{R}^d} e^{-|\xi|^2/4T} \rho(d\xi) < \infty . \tag{3}$$

Conversely, given a nonnegative Borel measure which satisfies (3) for some $T > 0$, the function

$$u(x,t) \equiv (g * \rho)(x,t)$$

is the unique nonnegative solution of (1) in S_T whose initial trace is ρ. Here

$$g(x,t) \equiv (4\pi t)^{-d/2} e^{-|x|^2/4t}$$

is the fundamental solution of (1). My objective in this Lecture is to outline the development of an analogous theory for the porous medium equation.

A function $u = u(x,t)$ is said to be a <u>continuous weak solution</u> of the porous medium equation

$$\frac{\partial u}{\partial t} = \Delta(u^m) \tag{4}$$

in S_T if it is continuous and nonnegative in S_T, and satisfies the integral identity

$$\iint_{\mathbb{R}^d \times (\tau_1,\tau_2)} u^m \Delta\psi + u\frac{\partial \psi}{\partial t} = \int_{\mathbb{R}^d} u\psi\Big|_{t=\tau_2} - \int_{\mathbb{R}^d} u\psi\Big|_{t=\tau_1}$$

for all τ_i such that $0 < \tau_1 < \tau_2 \leq T$ and for all $\psi \in C^{2,1}(S_T)$ such that $\psi(\cdot,t)$ has compact support for all $t \in [\tau_1,\tau_2]$.

It is proved in [AC1] that every continuous weak solution u of (4) posesses a unique initial trace ρ. Moreover, ρ must satisfy a growth condition analogous to (3) which limits the amount of mass that ρ can place at $|x| = \infty$. Specifically, there exists a constant $C = C(d,m) > 0$ such that

$$\int_{B_r(0)} \rho(dx) \leq C\{r^{\varkappa/(m-1)} T^{-1/(m-1)} + T^{d/2} u^{\varkappa/2}(0,T)\} \tag{5}$$

where

$$\varkappa = 2 + d(m-1) .$$

Roughly speaking, (5) means that, on average, $u(x,0)$ cannot grow faster than $|x|^{2/(m-1)}$ as $|x| \to \infty$, i.e., $v(x,0)$ cannot grow faster than $|x|^2$.

Introduce a norm on the space of nonnegative measures μ by

$$|||\mu||| \equiv \sup_{r \geq 1} r^{-\varkappa(m-1)} \mu(B_r(0)) .$$

Clearly (5) implies that $|||\rho||| < \infty$. Benilan, Crandall and Pierre [BCP] have shown that corresponding to every nonnegative measure ρ with $|||\rho||| < \infty$ there exists a function $W[\rho](x,t)$ which is a continuous weak solution of (4) in S_T for some $T > 0$ determined by $|||\rho|||$. Moreover, the initial trace of $W[\rho]$ is ρ. There still remains a uniqueness question, but this is settled by a very general uniqueness theorem due to Dahlberg and Kenig [DK1] which I will discuss later.

The growth estimate (5) is an easy consequence of the following Harnack-type inequality.

Theorem 1 [AC1]. Let u be a solution of (4) which is continuous in $\bar{S}_T = \mathbb{R}^{\alpha} \times [0,T]$ for some $T > 0$. There exists a constant $C = C(d,m) \in \mathbb{R}^+$ such that for any $\xi \in \mathbb{R}^d$

$$\int_{B_r(\xi)} u(x,0)\,dx \leq C\{r^{\varkappa/(m-1)} T^{-1/(m-1)} + T^{d/2} u^{\varkappa/2}(\xi,T)\} . \tag{6}$$

For any $q \in \mathbb{R}^+$, the quadratic pressure solution

$$\hat{W}(x,t;q) = \left(\frac{q|x|^2}{1 - \frac{qt}{t_o}}\right)^{1/(m-1)}$$

discussed in Lecture 1 is a continuous weak solution of (4) in $\bar{S}_T$ for any $T < \hat{T} = t_o/q$. Recall that $t_o = (m-1)/(2m\varkappa)$. Since

$$\int_{B_r(o)} \hat{W}(x,t;q)\,dx = q^{1/(m-1)} \omega_d \frac{m-1}{\varkappa} r^{\varkappa/(m-1)} = \frac{m-1}{\varkappa} t_o^{1/(m-1)} \omega_d \hat{T}^{-1/(m-1)} r^{\varkappa/(m-1)} ,$$

it follows that (6) is sharp.

The proof of Theorem 1 given in [AC1] is extremely technical and hard. I will sketch here a very elegent and conceptually simple proof due to Dahlberg and Kenig [DK2]. In all honesty it should be noted that the Dahlberg-Kenig proof relies on a uniqueness result due to Pierre [Pi] whose proof is technical and hard.

For simplicity take $\xi = 0$ and $m = 2$. By scaling, we can take $r = T = 1$ so that (6) becomes

$$\int_{B_1(0)} u(x,o)\,dx \leq C\{1+u^{1+\frac{d}{2}}(0,1)\} . \tag{7}$$

Moreover, by the maximum principle, it suffices to consider u's with $\operatorname{supp} u(\cdot,0) \subset B_1(0)$. Suppose that (7) is false, i.e., suppose there exists a sequence $\{u_k\}$ of solutions of (4) with $u_k \geq 0$, $\operatorname{supp} u_k(\cdot,0) \subset B_1(0)$, and

$$I_k \equiv \int_{B_1(0)} u_k(x,0)\,dx \geq k^{1+\frac{d}{2}} \{1+u_k^{1+\frac{d}{2}}(0,1)\} . \tag{8}$$

For any $\alpha_k \in \mathbb{R}^+$ the functions

$$w_k(x,t) \equiv \frac{1}{\alpha_k^2} u_k(\alpha_k x,t)$$

are also solutions of (4). Choose the sequence $\{\alpha_k\}$ so that $I_k = \alpha_k^{2+d}$. Note that $\operatorname{supp} w_k(\cdot,0) \subset B_{\alpha_k^{-1}}(0)$ and

$$\int_{\mathbb{R}^d} w_k(x,0)\,dx = 1 .$$

There exists a subsequence $w_k \to w$, where w is a continuous weak solution of (4) in S_1 with initial trace $\delta_o(d\mathbf{x})$. By Pierre's uniqueness theorem [Pi], w is the Barenblatt solution with mass 1. Now, in view of (8)

$$w_k(0,1) = \frac{u_k(0,1)}{\alpha_k^2} \leq \frac{1}{k} \to 0 \neq w(0,1) .$$

Therefore we have a contradiction and conclude that (7) holds. #

<u>Theorem 2</u> [AC1]. Suppose that u is a continuous weak solution of (4) in S_T for some $T>0$. Then there exists a unique nonnegative Borel measure ρ such that ρ is the initial trace of u and ρ satisfies (5).

The existence of an initial trace satisfying the growth condition (5) is an immediate consequence of the Harnack-type estimate (6) by a standard selection argument. The proof of the uniqueness of the trace involves additional labor.

The existence theorem of Benilan, Crandall, and Pierre [BCP] shows that (5) is not only necessary, but also sufficient.

<u>Theorem 3</u> [BCP]. Let ρ be a nonnegative Borel measure with $|||\rho||| < \infty$ and set

$$\ell = \limsup_{r\uparrow\infty} \sup_{R\geq r} r^{-\varkappa/(m-1)} \rho(B_r(0)) .$$

Then there exists a continuous weak solution $W[\rho](x,t)$ of (4) in S_T for

$$T = c(d,m)/\ell^{m-1} ,$$

where $c(d,m)$ is a positive constant depending only on d and m. Moreover,

$$W[\rho](x,t) \leq c(d,m)t^{-d/\varkappa}(1+|x|^2)^{1/(m-1)} |||\rho|||^{2/\varkappa}$$

in S_T.

There remains a rather subtle uniqueness question. Given a continuous weak solution u of (4) we determine a unique initial trace ρ satisfying (5). On the other hand, using Theorem 3 we can construct a continuous weak solution $W[\rho]$ with initial trace ρ. Does $W[\rho]=u$? The answer "yes" is provided by the following result of Dahlberg and Kenig.

<u>Theorem 4</u> [DK1]. Let u and w be continuous weak solutions of (4) in S_T. If

$$\lim_{t\downarrow 0} \{u(\cdot,t) - w(\cdot,t)\} = 0 \quad \text{in} \quad \mathcal{D}'(\mathbb{R}^d) \tag{9}$$

then $u=w$ everywhere in S_T.

In particular, (9) holds if u and w have the same initial trace. The proof of Theorem 4 uses Pierre's L^1 uniqueness theory [Pi] as well as the following pointwise estimates.

<u>Theorem 5</u> [DK1]. Let u be a continuous weak solution of (4) in S_T with initial trace ρ. Then

$$u(x,t) \leq C_t(u)\{1+|x|^2\}^{1/(m-1)}$$

where $C_t(u) = C_T \cdot C(u(0,T)) \cdot t^{d/\varkappa}$, and $C(u(0,T))$ is a constant which depends only on $u(0,T)$, d, and m.

Finally, Dahlberg and Kenig show that continuous weak solutions have all of the regularity properties which we have discussed.

<u>Theorem 6</u> [DK1]. If u is a continuous weak solution of (4) in S_T, then u is Holder continuous on any compact subset of S_T. Moreover, if $v = \frac{m}{m-1} u^{m-1}$ then

$$\Delta v \geq -\frac{k}{t} \quad \text{and} \quad \frac{\partial v}{\partial t} \geq -\frac{(m-1)kv}{t}$$

in $\mathcal{D}'(S_T)$ where $k = (m-1+\frac{2}{d})^{-1}$.

The initial trace theory for the "fast" diffusion case $m \in (0,1)$ has been worked out by Herrero and Pierre [HP]. They show the existence of a solution which is global in time for any $u_0 \in L^1_{loc}(\mathbb{R}^d)$, i.e., without any growth condition.

Lecture 7

LARGE TIME BEHAVIOR: INITIAL VALUE PROBLEMS. Consider the initial value problem

$$\frac{\partial u}{\partial t} = \Delta(u^m) \quad \text{in} \quad \mathbb{R}^d \times \mathbb{R}^+$$
$$u(\cdot,0) = u_0 \quad \text{in} \quad \mathbb{R}^d, \tag{1}$$

where we assume $u_0 \in C(\mathbb{R}^d) \cap L^1(\mathbb{R}^d) \cap L^2(\mathbb{R}^2)$, $u_0 \in [0,N]$ and $u_0 \not\equiv 0$. Set

$$M = \int_{\mathbb{R}^d} u_0 .$$

Sabinina [S] has shown that problem (1) has a unique generalized solution u with

$$\int_{\mathbb{R}^d} u(x,t)dx = M \quad \text{for all} \quad t \in \mathbb{R}^+ .$$

Moreover, Caffarelli and Friedman [CF3] have shown that u is Holder continuous uniformly for $(x,t) \in \mathbb{R}^d \times [\delta,\infty)$ for any $\delta > 0$.

In Lecture 3, I described a very precise result due to Vazquez [V1] on asymptotic behavior of solutions to (1) in the case $d = 1$. There is, at present, no complete extension of Vazquez's result for $d > 1$. The following result due to Friedman and Kamin [FK] is the best obtained so far.

Theorem 1 [FK]. Let u be the solution to (1) and let $U(\cdot,\cdot;M)$ be the Barenblatt solution with mass

$$M = \int_{\mathbb{R}^d} u_0 .$$

Then as $t \uparrow \infty$

$$t^{d/\varkappa}|u(x,t) - U(x,t;M)| \to 0 \qquad (\varkappa = d(m-1)+2)$$

uniformly with respect to x in any set of the form

$$\{(x,t) \in \mathbb{R}^d \times \mathbb{R}^+ : |x| \le Ct^{1/n}\}$$

for some constant $C > 0$.

LARGE TIME BEHAVIOR: BOUNDARY VALUE PROBLEMS. Let Ω be a bounded domain in $\mathbb{R}^d$ and consider the initial-boundary value problem

$$\frac{\partial u}{\partial t} = \Delta(u^m) \quad \text{in } \Omega\times\mathbb{R}^+$$

$$u = 0 \quad \text{in } \partial\Omega\times\mathbb{R}^d \tag{2}$$

$$u(\cdot,0) = u_o \quad \text{in } \Omega .$$

Here u_o is a given nonnegative function.

For the equation of heat conduction, $m=1$, it is well known that under appropriate conditions on u_o and Ω, $u(\cdot,t)\to 0$ as $t\to\infty$. To be more precise, let $0<\lambda_o<\lambda_1\leq\ldots$ denote the eigenvalues of the Laplace operator on Ω with Dirichlet boundary conditions, and let $\varphi_o(x)$ denote the eigenfunction corresponding to λ_o. We can take φ_o to be positive and normalize so that $\|\varphi_o\|_{L^2(\Omega)} = 1$. Then there exist positive constants K_1 and K_2 depending only on u_o and Ω such that

$$\|u(\cdot,t)\|_{L^2(\Omega)} \leq K_1 e^{-\lambda_o t} \quad \text{for } t\geq 0$$

and

$$\|e^{\lambda_o t} u(\cdot,t) - \langle\varphi_o,u_o\rangle\varphi_o\|_{L^2(\Omega)} \leq K_2 e^{-(\lambda_1-\lambda_o)t} \quad \text{for } t\geq 0 .$$

Here $\langle\cdot,\cdot\rangle$ is the usual scalar product on $L^2(\Omega)$.

I want to describe the analogous results for the case $m>1$. The main result is the estimate

$$\|(1+t)^{1/(m-1)}u(\cdot,t)-f\|_{L^\infty(\Omega)} \leq K(1+t)^{-1} , \tag{3}$$

where $K\in\mathbb{R}^+$ is a constant which depends only on d,m,u_o and Ω, and f is the unique nontrivial solution of the problem

$$\Delta(f^m) + \frac{1}{m-1} f = 0 \quad \text{in } \Omega$$

$$f = 0 \quad \text{on } \partial\Omega . \tag{4}$$

The estimate (3) is sharp.

To prove these results we must have some hypotheses. I assume:

(H_1) Ω is a bounded arcwise connected open set with compact closure whose boundary is compact and of class C^3.

(H_2) $u_o(x)$ is a nonnegative continuous function defined on $\bar\Omega$ such that $u_o=0$ on $\partial\Omega$ and $u_o^m\in C^1(\bar\Omega)$.

I will refer to these hypotheses collectively as (H). With some labor, the assumption that $\partial\Omega \in C^3$ can be replaced by $\partial\Omega \in C^{2+1/m}$.

The first thing to do is to establish existence and uniqueness for the transient problem (2) and the steady state problem (4). Let $Q_T = \Omega \times (0,T]$ and $S_T = \partial\Omega \times (0,T]$ for any $T \in \mathbb{R}^+$. A function $u: Q_T \to [0,\infty)$ is said to be a generalized solution of problem (2) in Q_T if:

(i) For each $(y,t) \in S_T$, $\lim_{Q_T \ni (x,t) \to (y,t)} u(x,t) = 0$.

(ii) u and ∇u^m are in $L^2(Q_T)$.

(iii) u satisfies

$$\int_{Q_T} \nabla\psi \cdot \nabla u^m - \psi_t u = \int_{\Omega} \psi(x,0) u_0$$

for all $\psi \in C^1(\bar{Q}_T)$ which vanish on S_T and $\Omega \times \{T\}$.

Proposition 1. [AP] (A) If (H) holds then problem (2) possesses a unique generalized solution in Q_T for any $T \in \mathbb{R}^+$.

(B) The comparison principle holds for generalized solutions of problem (2).

Proposition 1 is proved in [AP] by an approximation and regularization argument. Gilding and Peletier [GP2] have proved that the generalized solution of (1) belongs to $C(\bar{Q}_T)$ for every $T \in \mathbb{R}^+$.

Now consider the nonlinear eigenvalue problem

$$\begin{aligned} -\Delta u &= \lambda u^\delta \quad \text{in } \Omega \\ u &= 0 \quad \text{on } \partial\Omega \end{aligned} \tag{5}$$

where $\delta \in (0,1)$. Clearly problem (5) reduces to problem (3) if we set $u = f^m$, $\lambda = 1/(m-1)$, and $\delta = 1/m$. To ensure that u^δ is defined we consider only nonnegative solutions of (5).

Let $\bar{u} \in C^2(\Omega) \cap C(\bar{\Omega})$ be a solution of (5) for $\lambda = \bar{\lambda}$. We say that $\bar{u}$ is an eigenfunction of problem (5) corresponding to $\lambda = \bar{\lambda}$ if $\bar{u} \geq 0$ and $\bar{u} \not\equiv 0$. If $\bar{u}$ is an eigenfunction of problem (5) corresponding to $\bar{\lambda}$ then $u = (\lambda/\bar{\lambda})^{1/(1-\delta)}\bar{u}$ is an eigenfunction of problem (5) corresponding to λ. Thus to prove the existence of eigenfunctions it suffices to consider a single value of λ, say $\lambda = 1$.

Proposition 2 [AP]. If Ω satisfies (H_1) then for each $\lambda \in \mathbb{R}^+$ problem (5) has a unique eigenfunction u. Moreover, $u \in C^{2+\delta}(\bar{\Omega})$, $u > 0$ in Ω and $\frac{\partial u}{\partial \nu} < 0$ on $\partial\Omega$.

Here $\partial/\partial\nu$ denotes the derivative in the direction of the outer normal on $\partial\Omega$.

Sketch of proof. To prove the existence of an eigenfunction in $C^{2+\delta}(\bar{\Omega})$ for $\lambda = 1$ one can use a result due to Amann [A]. It suffices to establish the existence of a $v \in C^{2+\delta}(\bar{\Omega})$ such that

$$-\Delta v \geq v^{\delta} \quad \text{in } \Omega \text{ , } v \geq 0 \quad \text{on } \partial\Omega$$

and a function w such that $0 < w \leq v$ in Ω and

$$-\Delta w \leq w^{\delta} \quad \text{in } \Omega \text{ , } w = 0 \quad \text{on } \partial\Omega \text{ .}$$

To construct v , let Ω' be a smoothly bounded domain such that $\bar{\Omega} \subset \Omega'$ and consider the eigenvalue problem

$$-\Delta z = \mu z \quad \text{in } \Omega' \text{ , } z = 0 \quad \text{on } \partial\Omega' \text{ .} \tag{6}$$

Then $v = k\varphi_o$, where φ_o is the positive eigenfunction corresponding to the principle eigenvalue μ_o of (6) and $k = (\beta\mu_o)^{-1/(1-\delta)}$ with

$$\beta = \inf\{\varphi_o(x) : x \in \Omega\} \text{ .}$$

The construction of w is similar but more technical. See [AP] for details.

To prove uniqueness, suppose there are two eigenfunctions u_1 and u_2 corresponding to $\lambda = 1$, with $u_j > 0$ on Ω and $\partial u_j/\partial\nu < 0$ on $\partial\Omega$. Without loss of generality assume that $u_1 \not\geq u_2$ in Ω , i.e., $u_1 < u_2$ holds for at least some points of Ω . Define

$$\tau_o \equiv \sup\{\tau > 0 : \tau u_1 \not\geq u_2 \text{ in } \Omega\} \text{ .}$$

Since $u_1 \not\geq u_2$ and the u_i are continuous, we have $\tau_o > 1$. On the other hand, $\tau u_1 \geq u_2$ for all sufficiently large τ . Therefore $\tau_o \in (1,\infty)$. Set $z \equiv \tau_o u_1 - u_2$. Then $z \geq 0$ in Ω , $z = 0$ on $\partial\Omega$ and

$$-\Delta z = (\tau_o - \tau_o^{\delta})u_1^{\delta} + (\tau_o u_1)^{\delta} - u_2^{\delta} > 0 \quad \text{in } \Omega \text{ .}$$

By the strong maximum principle and the boundary point lemma [PW], $z > 0$ in Ω and $\frac{\partial z}{\partial\nu} < 0$ on $\partial\Omega$. Define $\tau_n \equiv \tau_o - 1/n$. There exist $x_n \in \Omega$ such that $\tau_n u_1(x_n) - u_2(x_n) < 0$. Since $\bar{\Omega}$ is compact there exists a subsequence $\{x_n\}$ such that $x_n \to x_o \in \bar{\Omega}$. If $x_o \in \Omega$ then $z(x_o) = 0$ which is a contradiction. Therefore $x_o \in \partial\Omega$. In view of (H_1) , if n is sufficiently large then there exists a sequence $z_n = z_n(x_n) \in \partial\Omega$ such that $d(x_n,\partial\Omega) = |x_n - z_n|$ and $z_n \to x_o$. By the theorem of the mean

$$0 < -\frac{\tau_n u_1(x_n)-u_2(x_n)}{d(x_n,\partial\Omega)} = \nabla\{\tau_n u_1(\tilde{x}_n)-u_2(\tilde{x}_n)\}\cdot \nu(z_n) ,$$

so that letting $n\to\infty$ we get $\partial z(x_o)/\partial\nu \geq 0$. Since this is again a contradiction, we conclude that $u_1 \geq u_2$ in Ω . The same argument with u_1 and u_2 interchanged gives $u_2 \geq u_1$. Therefore $u_1 = u_2$. #

Knowing the existence of u and f we can now give the formal statement of the main result.

Theorem 2 [AP]. If (H) holds then the solution of problem (2) satisfies (3).

The proof of Theorem 2 given in [AP] involves several rather technical comparison arguments involving various special selfsimilar solutions. It is relatively easy to show that there is a constant $\tau_1 > 0$ such that

$$0 \leq u(v,t) \leq (\tau_1+t)^{-1/(m-1)} f(x) .$$

However, it is much more difficult to obtain the corresponding lower bound since there is no reason to exclude initial functions u_o which have compact support in Ω . Hence we need the following very technical result.

Proposition 3 [AP]. Assume that (H) holds. If $u_o \not\equiv 0$ then there exists a $T \in [0,\infty)$ which depends only on d,m,u_o and Ω such that $u(x,T) > 0$ in $\Omega\times[T,\infty)$.

Note that if we drop (H_1) then it is possible to imagine domains where the time required to spread positivity is infinite, e.g. a chambered nautilus.

To conclude this lecture, I want to describe an application of Theorem 2 to the Gurten-MacCamy theory of "crowd avoiding" populations with Mathusian growth [GM]. The model equations are

$$\begin{aligned} &\rho_t = \Delta(\rho^m) + \mu\rho \quad \text{in } \Omega\times\mathbb{R}^+ \\ &\rho = 0 \text{ on } \partial\Omega\times\mathbb{R}^+ ,\ \rho = \rho_o \quad \text{in } \Omega\times\{0\} , \end{aligned} \tag{7}$$

where ρ represents population density. Assume the process is birth dominant, i.e., $\mu \in \mathbb{R}^+$. Then if $R(x,\tau) = u(x,t(\tau))$ is defined by

$$\rho(x,t) = e^{\mu t}u(x,t) \ , \ \tau = \frac{\gamma}{\mu}(e^{\mu t/\gamma} - 1) \qquad (\gamma = \frac{1}{m-1})$$

it follows that R is a solution of problem (2). Then applying Theorem 2 and translating the result back to the original variables leads to the following statement.

Theorem 3. [AP]. Assume Ω and ρ_o satisfy (H) and let ρ denote the solution of problem (7) for some $\mu \in \mathbb{R}^+$. There exists a constant C^* which depends only on d,m,μ,ρ_o and Ω such that

$$|\rho(x,t) - (\tfrac{\mu}{\gamma})^{\gamma} f(x)| \leq C^*(\tfrac{\mu}{\gamma})^{\gamma} f(x) e^{-\mu t/\gamma} \quad \text{in } \bar{\Omega}\times[0,\infty) .$$

Lecture 8

STABILIZATION. Consider the initial-boundary value problem

$$\begin{aligned} &u_t = (u^m)_{xx} + u(1-u)(u-a) \quad \text{in } (-L,L)\times\mathbb{R}^+ \\ &u(\pm L,\cdot) = 0 \quad \text{in } \mathbb{R}^+ \qquad\qquad (1) \\ &u(\cdot,0) = u_o \quad \text{in } (-L,L) \ , \end{aligned}$$

where $a \in (0,1/2)$ and u_o is a given nonnegative function. Problem (1) arises in the Gurten-MacCamy theory of crowd avoiding populations with a nonlinear growth law [GM]. This problem was studied in [ACP], where it was shown that every solution tends to an equilibrium solution as $t\to\infty$. Moreover, the structure of the equilibrium set was analyzed in detail in [ACP] and it was shown that for L sufficiently large it contains a continuum of solutions. In this lecture I want to present an alternative treatment of the basic stabilization result due to Langlais and Phillips [LP]. Their treatment has two advantages. It works in $\mathbb{R}^d$ for any $d>1$ and avoids the rather difficult compactness questions dealt with in [ACP].

On the formal level, the stabilization result for (1) follows from the existence of a functional which decreases on solutions of (1), i.e., a Liapunov function. Actually, one can find such a functional in a much more general setting. Let Ω be a bounded domain in $\mathbb{R}^d$ with smooth boundary. Consider the initial-boundary value problem

$$\begin{aligned} &\frac{\partial u}{\partial t} - \Delta\eta(u) = f(x,u) \quad \text{in } \Omega\times\mathbb{R}^+ \\ &u=0 \quad \text{on } \partial\Omega\times\mathbb{R}^+ \qquad\qquad (2) \\ &u(\cdot,0) = u_o \quad \text{on } \Omega \ , \end{aligned}$$

where $u_o \geq 0$ is a given function. Langlais and Phillips also allow nonzero boundary values, but I will not discuss that here. Throughout the lecture I will assume

$$\eta \in C^1[0,\infty)\cap C^3(0,\infty) \ , \ \eta(0) = 0 \ , \ \eta'(s)>0 \ \text{ for } \ s>0 \ . \qquad (3)$$

Formally, if we multiply both sides of the differential equation in (2) by $\partial\eta(u)/\partial t$, integrate by parts on $\Omega\times(t_1,t_2)$, and use the fact that $\partial u/\partial t = 0$ on $\partial\Omega$ we obtain

$$\int_\Omega \tfrac{1}{2}|\nabla\eta(u)|^2 - F(x,u)\Big|_{t=t_2} = \int_\Omega \tfrac{1}{2}|\nabla\eta(u)|^2 - F(x,u)\Big|_{t=t_1} - \iint_{\Omega\times(t_1,t_2)} \eta'(u)\left(\frac{\partial u}{\partial t}\right)^2 \qquad (4)$$

where

$$F(x,u) \equiv \int_0^u f(x,s)\eta'(s)ds .$$

Thus, in view of (3), the functional

$$L[u] \equiv \int_\Omega \tfrac{1}{2}|\nabla\eta(u)|^2 - F(x,u)$$

decreases on solutions to (2).

Since I will regard the solution of (2) as a flow in an appropriate Banach space, I will often suppress the space variables and write $u = u(t;u_0)$. The main result of [LP] concerns the relationship between the <u>ω-limit set</u>

$$\omega(u_0) \equiv \{w \in W^{1,2}(\Omega) \cap L^\infty(\Omega) : \text{there exists a sequence } t_n \to \infty \text{ with } \eta(u)(t_n,u_0) \to w \text{ in } L^2(\Omega) \text{ as } n \to \infty\}$$

and the equilibrium set $\mathcal{E}$ consisting of solutions w of the steady state problem

$$-\Delta w = f(x,\eta^{-1}(w)) \quad \text{in } \Omega$$

$$w = 0 \quad \text{on } \partial\Omega .$$

Under suitable hypothesis on f and u_0, the semiorbit $\{\eta(u)(t;u_0) : t \geq t_0\}$ is bounded in $W^{1,2}(\Omega)$ so that $\omega(u_0)$ is well-defined. Moreover,

$$\omega(u_0) \subset \mathcal{E} ,$$

i.e. $w \in \omega(u_0)$ implies that w is a steady state solution.

Let $Q_T = \Omega\times(0,T]$ and $S_T = \partial\Omega\times(0,T]$. Assume that

$$f \in C(\bar\Omega\times[0,\infty)) , \; f(x,0) = 0 , \; u_0 \geq 0 , \; u_0 \in L^\infty(\Omega) . \qquad (5)$$

A nonnegative function $u = u(t;u_0)$ is said to be a weak solution to (2) if, for every $T \in \mathbb{R}^+$, $u \in C([0,T];L^1(\Omega)) \cap L^\infty(Q_T)$ and

$$\int_\Omega u\psi\,\Big|_{t=T} - \iint_{Q_T} u\psi_t + \eta(u)\Delta\psi + f(x,u)\psi = \int_\Omega u_0\psi\,\Big|_{t=0} \qquad (6)$$

for every test function $\psi \in C^2(\bar Q_T)$ which vanishes on S_T.

The first result in [LP] is that if u_o and f are such that weak solutions of (2) are sufficiently regular then $\omega(u_o) \subset \mathcal{E}$.

Theorem 1. [LP] Assume that (5) holds and let u be a weak solution of (2). If $u \in L^\infty(Q_\infty)$ and

$$\frac{\partial \eta(u)}{\partial t} \in L^2(\Omega \times (t_o,\infty)) \ , \ \nabla \eta(u) \in L^\infty(t_o,\infty;L^2(\Omega)) \tag{7}$$

for some $t_o > 0$, then $\omega(u_o)$ is well-defined and $w \in \omega(u_o)$ implies $w \in \mathcal{E}$.

Sketch of proof. In view of (7) and the boundedness of u , $\omega(u_o)$ is well-defined. Suppose that

$$w = \lim_{n\to\infty} \eta(u)(t_n;u_o) \quad \text{in} \quad L^2(\Omega)$$

and define

$$U_n(x,s) \equiv \eta(u)(t_n+s;u_o) \quad \text{in} \quad \Omega \times (-1,1) \ .$$

I claim that $U_n \to w$ in $L^2(\Omega) \times (-1,1))$ as $n \to \infty$. Indeed, by the Schwarz inequality

$$\int_\Omega |\eta(u)(t_n+s;u_o) - \eta(u)(t_n;u_o)|^2 \le 2 \int_\Omega \int_{t_n-1}^{t_n+1} \{\frac{\partial \eta(u)}{\partial t}\}^2 \ ,$$

which implies

$$\|U_n - \eta(u)(t_n;u_o)\|_{L^2(\Omega \times (-1,1))} \le 2[\int_\Omega \int_{t_n-1}^\infty \{\frac{\partial \eta(u)}{\partial t}\}^2]^{1/2} \ .$$

The assertion follows since, by (7), the right hand side tends to zero as $n \to \infty$. For a subsequence, which we again call U_n , $U_n(x,s) \to w(x)$ a.e. in $\Omega \times (-1,1)$. Moreover,

$$u(t_n+s;u_o) \to \eta^{-1}(w) \quad \text{and} \quad f(x,u(t_n+s,u_o)) \to f(x,\eta^{-1}(w))$$

in $L^2(\Omega \times (-1,1))$ as $n \to \infty$ since all of the functions involved are uniformly bounded.

Let $\zeta \in C^2(\bar\Omega)$ be such that $\zeta = 0$ on $\partial\Omega$ and $p \in C^2(\mathbb{R})$ be such that $\operatorname{supp} p \subset [-1,1]$ and

$$\int_{-1}^{1} p\,ds = 1 \ .$$

Take $\psi_n(x,t) = p(t-t_n)\zeta(x)$ as a test function in the weak form of (2) with $T = t_n+1$ to get

$$\int_{t_n-1}^{t_n+1} \int_\Omega p'(t-t_n)\zeta u + \eta(u)p(t-t_n)\Delta\zeta + f(x,u)p(t-t_n)\zeta = 0 \ .$$

Now set $s = t - t_n$ and let $n \to \infty$. The result is

$$\int_{-1}^{1} p' \int_{\Omega} \eta^{-1} w\xi + \int_{-1}^{1} p \int_{\Omega} w\Delta\zeta + f(x,\eta^{-1}(w))\zeta = 0 .$$

Since

$$\int_{-1}^{1} p = 1 \quad \text{and} \quad \int_{-1}^{1} p' = 0$$

it follows that

$$\int_{\Omega} w\Delta\zeta + f(x,\eta^{-1}(w))\zeta = 0 . \qquad \#$$

The problem now is to find conditions which ensure that (7) holds for every weak solution of (2). Following Langlais and Phillips [LP] we assume:

(H_1) $\eta'(u)$ increases on $\mathbb{R}^+$ and there exists $\gamma \in \mathbb{R}^+$ such that

$$\eta'(u) < \gamma \frac{\eta(u)}{u} \text{ in } \mathbb{R}^+ .$$

(H_2) For each $M \in \mathbb{R}^+$, $f \in C^{\beta}(\bar{\Omega} \times [0,M])$ for some $\beta > 0$, and there exists $K = K(M) \in \mathbb{R}^+$ such that

$$|f(x,u)-f(x,v)| \leq K|u-v|$$

for all $x \in \bar{\Omega}$ and all $u,v \in [0,M]$.

These conditions are sufficients to ensure the existence of a unique bounded weak solution to (2) if u_o is bounded.

<u>Theorem 2</u>. [LP] If $(H_1),(H_2)$ and (5) hold then $\omega(u_o)$ is well defined and $\omega(u_o) \subset \mathcal{E}$.

To prove Theorem 2 it suffices to show that (H_1) and (H_2) imply that (7) holds and then apply Theorem 1. The key step in the proof that (H_1) and (H_2) imply (7) is an energy estimate for the regularized problem

$$\frac{\partial u}{\partial t} - \Delta\eta(u) = f(x,u-\epsilon) \text{ in } Q_\infty$$
$$u = \epsilon \text{ on } S_\infty , \; u(\cdot,0) = u_o + \epsilon \text{ on } \Omega \qquad (8)_\epsilon$$

where $\epsilon \in \mathbb{R}^+$ and $u_o \in C^\infty(\Omega)$ with $0 \leq u_o \leq M$. By the standard theory of quasilinear parabolic equations [LSU], for each $\epsilon \in \mathbb{R}^+$ and $T \in \mathbb{R}^+$ problem $(8)_\epsilon$ has a unique classical solution $u = u_\epsilon \in C^{2,1}(\bar{Q}_T)$ with

$$0 < \epsilon \leq u_\epsilon \leq M+\epsilon \text{ in } \bar{Q}_T . \qquad (9)$$

<u>Lemma</u>. [LP] There exists a constant $C(M) > 0$ independent of ϵ and T such that for all $T \geq 1$ and $\epsilon \in (0,1)$

$$\int_1^T \int_{\Omega} \{\frac{\partial\eta(u_\epsilon)}{\partial t}\}^2 + \frac{1}{2}\int_{\Omega} |\nabla\eta(u_\epsilon)|^2 \Big|_{t=T} \leq C(M) .$$

Sketch of proof. Set

$$N(u) \equiv \int_0^u \{\eta(s)-\eta(\epsilon)\}ds \quad \text{and} \quad F_\epsilon(x,u) \equiv \int_0^u f(x,s-\epsilon)\eta'(s)ds .$$

Multiply both side of equation $(8)_\epsilon$ by $\eta(u_\epsilon)-\eta(\epsilon)$ and integrate by parts over Q_1. Using the fact that $\eta(u_\epsilon)-\eta(\epsilon)=0$ on S_1, it follows that

$$\int_0^1 \int_\Omega |\nabla\eta(u_\epsilon)|^2 = \int_0^1 \int_\Omega f(x,u_\epsilon-\epsilon)\{\eta(u_\epsilon)-\eta(\epsilon)\} + \int_\Omega N(u_\epsilon)\Big|_{t=1} - N(u_\epsilon)\Big|_{t=0} .$$

In view of (9), the right hand side is bounded by a constant $C_1(M) \in \mathbb{R}^+$ independent of ϵ and T. Thus by the theorem of the mean, there exists a $t_\epsilon \in (0,1)$ such that

$$\int_\Omega |\nabla\eta(u_\epsilon)|^2\Big|_{t=t_\epsilon} \leq C_1(M) . \tag{10}$$

Now multiply both sides of equation $(8)_\epsilon$ by $\frac{\partial}{\partial t}\eta(u_\epsilon)$ and integrate by parts on $\Omega\times(t_\epsilon,T)$ to obtain

$$\int_{t_\epsilon}^T \int_\Omega \eta'(u_\epsilon)\left(\frac{\partial u_\epsilon}{\partial t}\right)^2 dxdt + \frac{1}{2}\int_\Omega |\nabla\eta(u_\epsilon)|^2\Big|_{t=T} = \frac{1}{2}\int_\Omega |\nabla\eta(u_\epsilon)|^2\Big|_{t=t_\epsilon}$$

$$+ \int_\Omega F_\epsilon(x,u_\epsilon)\Big|_{t=T} - F_\epsilon(x,u_\epsilon)\Big|_{t=t_\epsilon} .$$

In view of (H_2), (9) and (10), the right hand side is bounded by a constant $C_2(M) \in \mathbb{R}^+$ independent of T and ϵ. The assertion of the Lemma follows since, by (H_1), there exists a constant $C_3(M) \in \mathbb{R}^+$ such that $\eta'(u_\epsilon) \leq C_3(M)$. #

The proof of Theorem 2 from the Lemma involves an essentially standard limit argument.

It is easily checked that problem (1) satisfies all of the hypotheses of Theorem 2. Thus the stabilization theorem in [ACP] is a consequence of Theorem 2. In addition to the stabilization theorem, [ACP] contains a detailed description of the structure of the equilibrium set for problem (1) as a function of the parameter L and estimates for the basin of attraction of the stable equilibria. The structure of the equilibrium set for the analog of problem (1) in $\mathbb{R}^d$ is unknown.

References

[A] Amann, H. On the existence of positive solutions of nonlinear elliptic boundary value problems, Indiana Univ. Math. J., 21 (1971), 125-146.

[An] Angenent, S. Analyticity of the interface of the porous medium equation after the waiting time, Mathematical Institute, University of Leiden, Report No. 30, 1985.

[Ar1] Aronson, D.G. Regularity properties of flows through porous media, SIAM J. Appl. Math., 17 (1969), 461-467.

[Ar2] Aronson, D.G. Regularity properties of flows through porous media: a counter-example, SIAM J. Appl. Math., 19 (1970), 299-307.

[Ar3] Aronson, D.G. Regularity properties of flows through porous media: the interface, Arch. Rat. Mech. Anal., 37 (1970), 1-10.

[A4] Aronson, D.G. Nonlinear Diffusion Problems, in Free Boundary Problems: Theory and Applications, Vol. 1 (A. Fasano and M. Primicerio, editors), Research Notes in Mathematics 78, Pitman, London, 1983, 135-149.

[AB] Aronson D.G. and Benilan, Ph. Regularite des solutions de l'equation des milieux poreux dans $\mathbb{R}^N$, C.R. Acad. Sc. Paris, 288 (1979), 103-105.

[AC1] Aronson, D.G. and Caffarelli, L.A. The initial trace of a solution of the porous medium equation, Trans. Amer. Math. Soc., 280 (1983), 351-366.

[AC2] Aronson, D.G. and Caffarelli, L.A. Optimal regularity for one dimensional porous medium flow, in preparation.

[ACK] Aronson, D.G., Caffarelli, L.A. and Kamin, S. How an initially stationary interface begins to move in porous medium flow, SIAM J. Math. Anal., 14 (1983), 639-658.

[ACV] Aronson, D.G., Caffarelli, L.A. and Vazquez, J.L. Interfaces with a corner point in one-dimensional porous medium flow, Comm. Pure Appl. Math., 38 (1985), 375-404.

[ACP] Aronson, D.G., Crandall, M.G. and Peletier, L.A. Stabilization of solutions of a degenerate nonlinear diffusion problem, Nonlinear Anal. TMA, 6 (1982), 1001-1022.

[AG] Aronson, D.G. and Graveleau, J. In preparation.

[AP] Aronson, D.G. and Peletier, L.A. Large time behaviour of solutions of the porous medium equation in bounded domains, J. Diff. Eq., 39 (1981), 378-412.

[AV1] Aronson, D.G. and Vazquez, J.L. The porous medium equation as a finite speed approximation to a Hamilton-Jacobi equation, Inst. for Math. and its Applications Preprint Series 143, 1985; Analyse Non Lineaire, to appear.

[AV2] Aronson, D.G. and Vazquez, J.L. Eventual C^∞-regularity and concavity for flows in one dimensional porous media, in preparation.

[B1] Barenblatt, G.I. On some unsteady motions of a liquid or a gas in a porous medium, Prikl. Mat. Meh., 16 (1952), 67-78.

[B2] Barenblatt, G.I. Similarity, Self-Similarity, and Intermediate Asymptotics, Consultants Bureau, New York, 1979.

[Be] Benilan, Ph. A strong regularity L^p for solutions of the porous media equation, in Contributions to Nonlinear Partial Differential Equations (C. Bardos, A. Damlamian, J.I. Diaz and J. Hernandez editors), Research Notes in Math. 89, Pitman, London, 1983, 39-58.

[BBC] Benilan, Ph., Brezis, H. and Crandall, M.G. A semilinear elliptic equation in $L^1(\mathbb{R}^N)$, Ann. Scuola Norm. Sup. Pisa, 2 (1975), 523-555.

[BC] Benilan, Ph. and Crandall, M.G. The continuous dependence on φ of solutions of $u_t - \Delta\varphi(u) = 0$, Indiana Univ. Math. J., 30 (1981), 161-177.

[BCP] Benilan, Ph., Crandall, M.G. and Pierre, M. Solutions of the porous medium equation in $\mathbb{R}^N$ under optimal conditions on initial values, Indiana Univ. Math. J., 33 (1984), 51-87.

[BH] Berryman, J.G. and Holland, C.J. Stability of the separable solution for fast diffusion, Arch. Rat. Mech. Anal., 74 (1980), 279-288.

[BP] Bertsch, M. and Peletier, L.A. Porous medium type equations: An overview, Mathematical Institute, University of Leiden, Report No. 7, 1983.

[CF1] Caffarelli, L.A. and Friedman, A. Regularity of the free boundary for the one-dimensional flow of gas in a porous medium, Amer. J. Math., 101 (1979), 1193-1281.

[CF2] Caffarelli, L.A. and Friedman, A. Continuity of the density of a gas flow in a porous medium, Trans. Amer. Math. Soc., 252 (1979), 99-113.

[CF3] Caffarelli, L.A. and Friedman, A. Regularity of the free boundary of a gas flow in an n-dimensional porous medium, Indiana Univ. Math. J., 29 (1980), 361-391.

[CVW] Caffarelli, L.A., Vazquez, J.L. and Wolanski, N.I. Lipschitz continuity of solutions and interfaces of the N-dimensional porous medium equation, Inst. for Math. and its Applications. Preprint Series 191, 1985.

[CW] Caffarelli, L.A. and Wolanski, N.I. In preparation.

[CS] Chipot, M. and Sideris, T.S. An upper bound for the waiting time for nonlinear degenerate parabolic equations, Trans. Amer. Math. Soc., 288 (1985), 423-427.

[CL] Crandall, M.G. and Lions, P.L. Viscosity solutions of Hamilton-Jacobi equations, Trans. Amer. Math. Soc., 277 (1983), 1-42.

[D] Darcy, H. Les Fontaines Publiques de la Ville de Dijon, V. Dalmont, Paris, 1856, 305-311.

[DiB] DiBenedetto, E. Regularity results for the porous medium equation, Ann. Mat. Pura Appl., 121 (1979), 249-262.

[DK1] Dahlberg, B.E.J. and Kenig, C.E. Nonnegative solutions of the porous medium equation, Comm. PDE, 9 (1984), 409-437.

[DK2] Dahlberg, B.E.J. and Kenig, C.E. Nonlinear filtration, preprint.

[FK] Friedman, A. and Kamin, S. The asymptotic behavior of gas in an n-dimensional porous medium, Trans. Amer. Math.Soc., 262 (1980), 551-563.

[G] Gilding, B.H. Holder continuity of solutions of parabolic equations, J. London Math. Soc., 13 (1976), 103-106.

[GP1] Gilding, B.H. and Peletier, L.A. On a class of similarity solutions of the porous medium equation, J. Math. Anal. Appl., 55 (1976), 351-364: II, 57 (1977), 522-538.

[GP2] Gilding, B.H. and Peletier, L.A. Continuity of solutions of the porous medium equation, Ann. Scuola Norm. Sup. Pisa, 8 (1981), 659-675.

[Gr] Graveleau, J. Personel communication, 1973.

[GM] Gurtin, M.E. and MacCamy, R.C. On the diffusion of biological populations, Math. Biosc., 33 (1977), 35-49.

[HP] Herrero, M.A. and M. Pierre. The Cauchy problem for $u_t = \Delta u^m$ when $0 < m < 1$, Trans. Amer. Math. Soc., 291 (1985), 145-158.

[HK] Hollig, K. and Kreiss, H.O. C^∞-regularity for the porous medium equation, Univ. of Wisconsin-Madison, Computer Science Dept., Technical Report No. 600 1985.

[K] Kalashnikov, A.S. The Cauchy problem in a class of growing functions for equations of unsteady filration type, Vest. Mosk. Univ. Ser. Mat. Mech., 6 (1963), 17-27.

[Ka] Kamenomostskaya, S. (Kamin). The asymptotic behaviour of the solution of the filtration equation, Israel J. Math., 14 (1973), 76-78.

[Kn] Knerr, B.F. The porous medium equation in one dimension, Trans. Amer. Math. Soc., 234 (1977), 381-415.

[Kr] Kruzhkov, S.N. Results on the character of the regularity of solutions of parabolic equations and some of their applications, Math. Notes, 6 (1969), 517-523.

[LSU] Ladyzhenskaya, O.A., Solonnikov, V.A. and Ural'ceva, N.N. Linear and Quasilinear Equations of Parabolic Type, Transl. Math. Monographs 23, Amer. Math. Soc., Providence, 1968.

[LP] Langlais, M. and Phillips, D. Stabilization of solutions of nonlinear and degenerate evolution equations, Nonlinear Anal. TMA, 9 (1985), 321-333.

[L] Lions, P.L. Generalized Solutions of Hamilton-Jacobi Equations, Research Notes in Math., 69, Pitman, London, 1982.

[LSV] Lions, P.L., Souganidis, P.E. and Vazquez, J.L. In preparation.

[M] Muskat, M. The Flow of Homogeneous Fluids Through Porous Media, McGraw-Hill, New York, 1937.

[O] Oleinik, O.A. On some degenerate quasilinear parabolic equations, Seminari dell' Istituto Nazionale di Alta Matematica 1962-63, Oderisi, Gubbio, 1964, 355-371.

[OKC] Oleinik, O.A., Kalashnikov, A.S. and Chzhou Yui-Lin. The Cauchy problem and boundary problems for equations of the type of unsteady filtration, Izv. Akad. Nauk SSSR Ser. Mat., 22 (1958), 667-704.

[P] Peletier, L.A. The porous medium equation, in Applications of Nonlinear Analysis in the Physical Sciences (H. Amann, N. Bazley and K. Kirchgassner editors), Pitman, London, 1981, 229-241.

[Pi] Pierre, M. Uniqueness of the stolution of $u_t - \Delta\varphi(u) = 0$ with initial datum a measure, Nonlinear Anal. TMA, 6 (1982), 175-187.

[PW] Protter, M.H. and Weinberger, H.F. Maximum Principles in Differential Equations Prentice-Hall, Englewood Cliffs, 1967.

[S] Sabinina, E.S. On the Cauchy problem for the equation of nonstationary gas filtration in several space variables, Dokl. Akad. Nauk SSSR, 136 (1961), 1034-1037.

[TM] Tomoeda, K. and Mimura M. Numerical approximations to interface curves for a porous media equation, Hiroshima Math. J., 13 (1983), 273-294.

[V1] Vazquez, J.L. Asymptotic behaviour and propagation properties of the one-dimensional flow of gas in a porous medium, Trans. Amer. Math. Soc., 277 (1983), 507-527.

[V2] Vazquez, J.L. The interface of one-dimensional flows in porous media, Trans. Amer. Math. Soc., 285 (1984), 717-737.

[V3] Vazquez, J.L. Behaviour of the velocity of one-dimensional flows in porous media, Trans. Amer. Math. Soc., 286 (1984), 787-802.

[W] Widder, D.V. Positive temperature on an infinite rod, Trans. Amer. Math. Soc., 55 (1944), 85-95.

[ZR] Zeldovich, Ya.B. and Raizer, Yu.P. Physics of Shock-waves and High-temperature Hydrodynamic Phenomena Vol. II, Academic Press, New York, 1966.

QUALITATIVE METHODS FOR NONLINEAR DIFFUSION EQUATIONS

Jesus Hernàndez

INTRODUCTION

A considerable effort has been devoted during the last years to the study of nonlinear diffusion equations or reaction-diffusion systems. A typical example is the following semilinear parabolic problem

$$(0.1)\qquad \begin{aligned} u_t - a\,\Delta u &= f(u,v) \quad \text{in } \Omega \times (0,T) \;, \\ v_t - b\,\Delta v &= g(u,v) \quad \text{in } \Omega \times (0,T) \;, \end{aligned}$$

where Ω is a smooth not necessarily bounded domain in $\mathbb{R}^N$, $a,b,T > 0$ are real numbers and f and g are sufficiently smooth functions. Of course, the above system should be complemented with suitable boundary and initial conditions. These systems, where the interaction between diffusion (measured in (0.1) by the diffusion coefficients $a, b > 0$) and the reaction terms f and g gives rise to an extremely rich phenomenology, arise in many applications, giving qualitative models for a number of physical and biological phenomena. We can mention, among many others, chemical reactions (e.g., Brusselator), combustion theory, population dynamics (predator-prey interactions, competition, etc.), nerve conduction (Fitzhugh-Nagumo system), morphogenesis (activator-inhibitor interaction), superconductivity, etc. Cf. the book by Smoller [106] for more information and references (cf. also the lecture notes by Fife [54]).

One of the main mathematical problems posed by reaction-diffusion systems is the asymptotic behavior of solutions, i.e. to investigate what happens with solutions when the time t goes to infinity. (Clearly, this implies the existence and uniqueness of solutions). From this point of view it is particularly interesting to study the existence and (maybe) multiplicity of solutions of the stationary problem associated to (0.1), namely the elliptic system

$$
(0.2) \qquad \begin{aligned} - a\, \Delta u &= f(u,v) \quad \text{in } \Omega \ , \\ - b\, \Delta v &= g(u,v) \quad \text{in } \Omega \ , \end{aligned}
$$

together with the corresponding boundary conditions. Indeed, in many interesting cases, the solutions of the evolution problem converge to one of the solutions (maybe unique) of (0.2). Therefore, the study of the stability of stationary solutions is one of the most important problems in the theory. This is not the only possibility, and the reader can found a good survey of many possible situations (travelling waves, spiral patterns, λ-ω systems, periodic solutions, etc.) in a paper by Fife [55]. However, a great deal of work has been devoted to the stability of solutions of systems as (0.2). It is well-known that exchanges of stability when a parameter varies are closely related with bifurcation phenomena, i.e., with changes in the structure of the solution set of (0.2) when the parameter crosses a critical value. We do not intend to treat this question here, but the reader can found some references in the Comments.

Our aim in these notes is to show how some methods of Nonlinear Functional Analysis can be applied to the study of existence and multiplicity of some systems like (0.2). We do not pretend to give an exhaustive survey of the different mathematical tools, and a similar consideration can be made concerning the variety of problems we consider. On the contrary, our intention is to exhibit how some mathematical tools (as sub and supersolutions, degree theory, critical point theory, etc.) work in a few examples which seem interesting and useful for further extensions. However, this overview of qualitative methods (comparison methods, topological methods, variational methods) is by no means complete: to mention only two important instances, local bifurcation methods (as, e.g., the Liapunov-Schmidt method [34], [66], [88]) and Conley index ([38], [106]) are not considered. On the other hand, it is very well-known that the methods in these notes have other interesting applications: for example, comparison arguments involving sub and supersolutions (cf., e.g., [78], [79], [62], [86]), topological degree [99], local inversion techniques [43] and the fixed point index [10] have applications in the study of the stability of steady-state solutions, and sub and supersolutions [56] and degree theory [21] have been used in the study of travelling waves.

A particular attention is payed to the case of a single equation. Of course, it seems reasonable to deal with this case before to treat full systems. But there is also another reason for that: by using a decoupling technique, some systems can be reduced to a single equation with a nonlocal perturbation, and some of the proofs are very similar to the case of a single equation. Cf. Comments for more details.

Very often, we are only interested in positive solutions when dealing with problems arising in physical applications. This interest shall appear in several ways along these notes, in particular we will use positivity arguments concerning the natural order in the corresponding function spaces and some related tools, as the Maximum Principle. We also emphasize the use of a cone of positive functions, replacing then the topological degree by the fixed point index. Cf. the excellent survey by Amman [10] (and [74]) for a very complete exposition of this kind of arguments.

The first chapter contains an exposition of the method of sub and supersolutions and some applications. Section 1.1 creats the case of a semilinear elliptic equation, and the method is applied in Section 1.2 to the nonlinear eigenvalue problem

$$(0.3)\qquad \begin{aligned} -\Delta u + f(u) &= \lambda u \quad \text{in } \Omega \,, \\ u &= 0 \quad \text{on } \partial\Omega \,, \end{aligned}$$

where λ is a real parameter and f behaves, roughly speaking, as u^3, to obtain existence of positive solutions. A uniqueness theorem is also given. Section 1.3 includes an extension of the method to systems without monotonicity assumptions, and some applications.

Chapter 2 is devoted to topological methods, especially the Leray-Schauder degree and some of its applications. Section 2.1 gives a brief survey (without proofs) of the main properties of the degree, which are applied in Section 2.2 to obtain the exact number of solutions of (0.3) for some values of λ . Section 2.3 includes the global bifurcation theorem by Rabinowitz, and some applications, and Section 2.4 treats the case without bifurcation. The remainder of this chapter deals with positive solutions: Section 2.5 gives some global bifurcation theorems for positive solutions, Section 2.6 contains a brief survey of the main properties of the fixed point index, and in Section 2.7 a global existence theorem for positive solutions is used, together with *a priori* estimates, to prove existence of positive solutions of a system arising in chemical reactions.

Two different methods are introduced in Chapter 3. The model example (0.3) is revisited, this time by using a continuation method based on a local inversion theorem by Crandall-Rabinowitz. A few variants of (0.3) are also considered. Some indications about the application of similar methods to the study of "bending points" in the case without bifurcation are given in Section 3.2. Finally, in Section 3.3 we use variational methods to improve under additional hypotheses the results in Section 3.2.

These notes were prepared while the author was visiting the Istituto Matematico U. Dini at the University of Florence. He wants to thank the staff there, and in particular A. Fasano, M. Primicerio and M. Ughi for their invaluable friendship and hospitality.

I. THE METHOD OF SUB AND SUPERSOLUTIONS AND APPLICATIONS

This chapter is mainly concerned with the method of sub and supersolutions and some of its applications.

The first paragraph treats the case of a nonlinear elliptic equation. Even if these results are very well-known, we find convenient to treat them here with some detail not only because they are the best illustration for the general case of systems, but also because, as it was indicated in the Introduction, existence and multiplicity results for some systems can be obtained by reformulating the problems in such a way that they become mild perturbations of the nonlinear eigenvalue problem we consider in Section 1.2. For this problem we study existence and uniqueness of positive solutions, together with some interesting additional qualitative properties, mostly concerning the smoothness of the branch of positive solutions we obtain. Finally, the last paragraph considers the extension of the method of sub and supersolutions to systems, and includes a rather general existence theorem. Some applications are also given.

1.1. AN EXISTENCE THEOREM FOR SEMILINEAR ELLIPTIC EQUATIONS.

Our main interest here is to show how the method of sub and supersolutions works in a concrete but very illustrative instance, namely the case of a mildly semilinear elliptic equation. This method, which is in some sense reminiscent of Perron's method for solving Dirichlet's problem by using subharmonic functions, not only gives an existence proof but also provides an iterative scheme to approximate solutions by sequences converging monotonically to these solutions.

Now we consider the nonlinear elliptic problem

$$(1.1) \qquad -\Delta u = f(x,u) \quad \text{in } \Omega$$

$$(1.2) \qquad u = 0 \quad \text{on } \partial\Omega ,$$

where Ω is, as in all which follows, a connected bounded open subset of $\mathbb{R}^N$ $(N \geq 1)$ with a very smooth boundary $\partial\Omega$; in particular, we assume that $\partial\Omega$ is $C^{2,\alpha}$, $0 < \alpha < 1$, but the reader may think about $\partial\Omega$ as being C^3 or even C^∞.

We also assume that the function $f : \bar{\Omega} \times \mathbb{R} \to \mathbb{R}$ satisfies

(1.3) $\quad f \in C^1(\bar{\Omega} \times \mathbb{R})$.

(1.4) $\quad$ For any $x \in \bar{\Omega}$, $f(x,u)$ is increasing in u .

Remark 1.1: The assumption (1.4) can be weakened, it is sufficient that there exists $M > 0$ such that $\left|\frac{\partial f(x,u)}{\partial u}\right| \leq M$. Indeed, in this case problem (1.1), (1.2) is equivalent to the problem

$$\begin{aligned} -\Delta u + M u &= \tilde{f}(x,u) \quad \text{in } \Omega \ , \\ u &= 0 \qquad\quad\ \ \partial\Omega \ , \end{aligned}$$

where $\tilde{f}(x,u) = f(x,u) + M u$ satisfies (1.3), (1.4). Moreover, we shall see below that (1.4) should only be satisfied on some bounded interval in u , and then (1.4) will follow from (1.3) (Cf. Remark 1.5). On the other hand, (1.3) can also be replaced (cf. [2],[10]) by the more general condition that f is locally C^α, $0 < \alpha < 1$, and there is a $M > 0$ such that (1.4) is satisfied for $f(x,u) + M u$ instead of f .

First, we give the definition of sub and supersolutions for our problem.

Definition: A sufficiently smooth function (i.e., in $C^2(\Omega) \cap C(\bar{\Omega})$) u_o (resp. u^o) is called a subsolution (resp. a supersolution) of (1.1) (1.2) if we have

(1.5) $\quad -\Delta u_o - f(x,u_o) \leq 0 \leq -\Delta u^o - f(x,u^o) \quad \text{in } \Omega$,

(1.6) $\quad u_o \leq 0 \leq u^o \quad \text{on } \partial\Omega$,

where inequalities are to be understood pointwise.

Our aim is to prove the following existence result.

Theorem 1.1: Let u_o (resp. u^o) be a subsolution (resp. a supersolution) of (1.1) (1.2) such that $u_o \le u^o$ (in Ω). Then there exist $\underline{u}$ and $\bar{u}$ solutions of (1.1) (1.2) satisfying $u_o \le \underline{u} \le \bar{u} \le u^o$. Moreover, $\underline{u}$ (resp. $\bar{u}$) is minimal (resp. maximal) in the sense that if z is a solution such that $u_o \le z \le u^o$, then $\underline{u} \le z \le \bar{u}$.

A few remarks seem to be in order before giving the proof of Theorem 1.1. First, this is not the most general version of this kind of existence results, it is even the contrary, so as to say. Our main interest here is to demonstrate how the method works, and it seems convenient from this point of view to get rid of irrelevant technical difficulties. In fact, the theorem can be extended to much more general differential operators and boundary conditions (Cf. Comments). There is also an "abstract" version in ordered Banach spaces (cf. [6],[10]).

First, let us point out that both the minimal and the maximal solutions $\underline{u}$ and $\bar{u}$ may coincide, and this is obviously the case when the solution is unique. (By the way, here solution means classical solution, i.e., a function in $C^2(\Omega) \cap C(\bar{\Omega})$). But here, contrarily to other different situations (as, e.g., Banach's fixed point theorem), Theorem 1.1 has no implications concerning uniqueness, which should be proved (when there is !) by a different argument, exploiting the particular features of the corresponding problem. Cf. below, e.g., Theorem 1.3.

On the other hand, this method does not provide all the solutions of (1.1) (1.2). One way to show this is to exhibit an equation with three or more solutions. But there are also quite general, and interesting, results due to Sattinger [100],[102] and Ammann [10] saying that if the solution obtained in Theorem 1.1 is unique, then it is stable or, otherwise stated, that an unstable solution cannot be obtained by using our iterative scheme. Cf. Comments.

Remark 1.2: The fact that u_o is a subsolution and u^o a supersolution does not imply $u_o \le u^o$. As it is obvious that any solution is also a subsolution (and a supersolution), any problem having more than one solution shows that this is not the case. But it is not difficult (cf., e.g., below paragraph 1.2) to find examples of strict subsolutions above a (may be strict) supersolution. (Here strict means strict inequality somewhere in (1.5) (1.6)).

Remak 1.3: The theorem is false for a subsolution u_o and a supersolution u^o such that $u_o > u^o$ as shows the counterexample in [10, pp. 653-654]. Moreover, in this case, the problem has no solution at all.

Theorem 1.1 is proved by reducing the problem of finding solutions of (1.1) (1.2) to the equivalent problem of showing the existence of fixed points of a suitable nonlinear operator T defined on a function space adequately choosen. What we intend to show is that in this case we should take care not only of the topological (continuity, compactness) properties of T but also use the fact that T preserves the natural order in the corresponding function space.

Now we define our nonlinear operator T in such a way that its fixed points are (classical) solutions of (1.1) (1.2). For that, if Ω is a $C^{2,\alpha}$ domain with $0<\alpha<1$ we define a nonlinear operator

$$T : C^{1,\alpha}(\bar{\Omega}) \longrightarrow C^{1,\alpha}(\bar{\Omega}) ,$$

$$u \longrightarrow Tu ,$$

where, for $u \in C^{1,\alpha}(\bar{\Omega})$, Tu is the (unique) solution of the linear problem

$$(1.7) \qquad - \Delta Tu = f(x,u) \quad \text{in } \Omega ,$$

$$(1.8) \qquad Tu = 0 \quad \text{on } \partial\Omega .$$

(We employ the usual notations for the spaces of Hölder continuous functions, $C^{k,\alpha}(\bar{\Omega})$, $k \geq 0$ integer, $0<\alpha<1$, and Sobolev spaces $W^{k,p}(\Omega)$, $k > 0$, $1 \leq p < \infty$, and the corresponding norms; we also use the space $L^{\infty}(\Omega)$ and its norm. Cf. [58]).

First, the operator T is well-defined. Indeed, if $u \in C^{1,\alpha}(\bar{\Omega})$, $f(x,u) \in C^{1}(\bar{\Omega})$ and by the well-known C^{α} theory for linear equations

(Schauder Theory) (cf., e.g. [58],[75]) there is a unique solution $Tu \in C^{2,\alpha}(\bar{\Omega})$ for (1.7) (1.8). Moreover, it is an easy task to prove, by using Schauder's estimates, that T is compact, i.e., that T is continuous and that, for B bounded, $T(B)$ is relatively compact. We shall return after to the links between order, compactness and the choice of the function space: for the moment, let us say only that, for this problem, it is still possible to work in $C(\bar{\Omega})$, $C^{\alpha}(\bar{\Omega})$ or even $L^2(\Omega)$.

The following results, which follow easily from the Maximum Principle, show that T behaves "nicely" with respect to the natural order in the function space. Here $u \le v$ means $u(x) \le v(x)$ for any $x \in \Omega$.

<u>Lemma 1.1</u>: T is order-preserving, i.e., $u \le v$ implies $Tu \le Tv$.

<u>Proof</u>: From (1.7) (1.8) for u and v it follows by (1.4)

$$- \Delta(Tv - Tu) = f(x,v) - f(x,u) \ge 0 \quad \text{in } \Omega \ ,$$

$$Tv - Tu = 0 \quad \text{on } \partial\Omega \ ,$$

and by the Maximum Principle, $Tv - Tu \ge 0$.

<u>Lemma 1.2</u>: If u_o is a subsolution then $u_o \le Tu_o$.

<u>Proof</u>: By the definition of T

$$- \Delta Tu_o = f(x,u_o) \quad \text{in } \Omega \ ,$$

$$Tu_o = 0 \quad \text{on } \partial\Omega \ ,$$

and this, together with (1.5) (1.6) gives

$$- \Delta(Tu_o - u_o) = f(x,u_o) + \Delta u_o \ge 0 \quad \text{in } \Omega \ ,$$

$$Tu_o - u_o \ge 0 \quad \text{on } \partial\Omega \ ,$$

and again by the Maximum Principle, $Tu_o - u_o \ge 0$.

<u>Lemma 1.3</u>: If u^o ia a supersolution, then $u^o \ge Tu^o$.

Proof: As for Lemma 1.2.

Proof of Theorem 1.1: We define by iteration the sequences

$$u_1 = Tu_o, \quad \ldots, \quad u_n = Tu_{n-1}, \ldots$$

$$u^1 = Tu^o, \quad \ldots, \quad u^n = Tu^{n-1}, \ldots$$

It follows from Lemmas 1.1-1.3 that u_n is increasing, u^n is decreasing, and

$$u_o \le u_1 \le \ldots \le u_n \le \ldots \le u^n \le \ldots \le u^1 \le u^o \quad . \tag{1.9}$$

Hence, from (1.2) we obtain

$$\|u_n\|_{L^\infty(\Omega)} \le c \qquad \|u^n\|_{L^\infty(\Omega)} \le c \tag{1.10}$$

where c denotes (as in all which follows, if not otherwise stated) a constant depending on Ω, N, f, etc. but not on n. Now from (1.10) and the L^p estimates of Agmon-Douglis-Nirenberg [1] we have, for any $1 \le p < \infty$

$$\|u_n\|_{W^{2,p}(\Omega)} = \|Tu_{n-1}\|_{W^{2,p}(\Omega)} \le c\,\|f(x,u_{n-1})\|_{L^p(\Omega)} \le c$$

and analogously

$$\|u^n\|_{W^{2,p}(\Omega)} \le c \quad .$$

Hence there exist two subsequences, which we again denote by u_n and u^n, such that for $p > 1$

$$u_n \longrightarrow \underline{u} \quad \text{in} \quad W^{2,p}(\Omega) \quad \text{weakly}$$

$$u^n \longrightarrow \bar{u} \quad \text{in} \quad W^{2,p}(\Omega) \quad \text{weakly}$$

and by Morrey's Lemma (the immersion $W^{2,p}(\Omega) \hookrightarrow C^{1,\delta}(\bar{\Omega})$ is compact for $p > N$, $\delta = 1 - \frac{N}{p}$)

$$u_n \longrightarrow \underline{u} \quad \text{in} \quad C^{1,\alpha}(\bar{\Omega})$$

$$u^n \longrightarrow \bar{u} \quad \text{in} \quad C^{1,\alpha}(\bar{\Omega}) \quad .$$

(Actually they converge in $C^{1,\delta}(\bar{\Omega})$ for any $0 < \delta < 1$). On the other hand, the monotonicity in (1.9) implies that the full sequences also converge. Since T is continuous

$$\lim T u_{n-1} = \lim u_n = \underline{u} = T\underline{u}$$

and hence $\underline{u}$ is a fixed point of T. We prove in the same way that $\bar{u} = T\bar{u}$, and it follows immediately from (1.9) that $\underline{u} \le \bar{u}$.

Now, let z be a solution such that $u_o \le z \le u^o$. Clearly $z = Tz$, and by Lemmas 1.1-1.3, $u_n \le z \le u^n$. Hence $\underline{u} \le z \le \bar{u}$, and this ends the proof.

Remark 1.4: The monotone sequences u_n and u^n in the proof of Theorem 1.1 converge, respectively, to $\underline{u}$ and $\bar{u}$ in $C^{1,\alpha}(\bar{\Omega})$. In fact, they converge in $C^{2,\alpha}(\bar{\Omega})$. From

$$- \Delta(T\underline{u} - Tu_n) = f(x,\underline{u}) - f(x,u_n) \quad \text{in} \quad \Omega \quad ,$$

$$T\underline{u} - Tu_n = 0 \qquad\qquad \text{on} \quad \partial\Omega \quad ,$$

it follows by Schauder's estimates

$$\|T\underline{u} - Tu_n\|_{2,\alpha} \le c \, \|f(x,\underline{u}) - f(x,u_n)\|_\alpha \quad .$$

Since $u_n \longrightarrow \underline{u}$ in $C^{1,\alpha}(\bar{\Omega})$, the result follows easily.

Remark 1.5: The proof of Theorem 1.1 shows clearly that all we need is (1.4) on the bounded interval $[\min u_o, \max u^o]$, and this is an easy corollary of (1.3). (In fact, we can assume f locally Lipschitz in u). Hence (1.4) is superfluous. This clarifies the observations in Remark 1.1 by showing that we do not need to assume f increasing.

This makes a big difference with respect to the case of systems (cf. Section 1.3).

It is clear that the method relies, in an essential way, on the Maximum Principle from one side (order properties) and on the classical C^{α} (Schauder Theory, cf. [58],[75]) and L^p (Agmon-Douglis-Nirenberg [1]) regularity results on the other. This implies that in the case of elliptic equations, its scope is reduced to second order differential operators, roughly speaking. Cf. the Comments for some of the possible extensions of the method.

1.2. POSITIVE SOLUTIONS FOR A NONLINEAR EIGENVALUE PROBLEM.

In this section it is shown how the ideas of the method of sub and supersolutions can be applied to the study of the existence of positive solutions of the nonlinear eigenvalue problem

(1.11) $$-\Delta u + f(u) = \lambda u \quad \text{in} \quad \Omega \ ,$$

(1.12) $$u = 0 \quad \text{on} \quad \partial\Omega \ ,$$

where Ω is a smooth bounded domain in $\mathbb{R}^N$, λ is a real parameter, and $f : \mathbb{R} \longrightarrow \mathbb{R}$ satisfies the assumptions

(1.13) f is C^2, increasing, and $f(0) = f'(0) = 0$,

(1.14) $\dfrac{f(u)}{u}$ is strictly increasing for $u > 0$ and strictly decreasing for $u < 0$,

(1.15) $$\lim_{|u| \to +\infty} \frac{f(u)}{u} = +\infty \ .$$

Problems of this kind arise in the theory of nuclear reactors (cf. [36],[108]) and have been studied by several authors (cf. [36], [108],[61],[20],[13]).

First, let us introduce a very useful notation. If the function ρ is smooth enough (this means $C(\bar{\Omega})$ or even $L^{\infty}(\Omega)$) , the eigenvalue problem

(1.16) $$- \Delta w + \rho w = \lambda w \quad \text{in} \quad \Omega$$

(1.17) $$w = 0 \quad \text{on} \quad \partial\Omega$$

posseses an infinite sequence of eigenvalues $\lambda_n(\rho)$, which can be ordered according to their multiplicity, going to $+\infty$ when $n \to +\infty$, and depending in a increasing continuous way on ρ (cf. [41],[106]). The classical book by Courant-Hilbert [41] is a good reference for all which concerns the properties of eigenvalues and eigenfunctions, in particular the continuous and monotone dependence on the coefficient ρ and the domain Ω and the variational characterization of the eigenvalues, etc, which will be very useful in the following.

In the particular case $\rho \equiv 0$, we will use the notation $\lambda_n = \lambda_n(0)$ for the eigenvalues of

(1.18) $$- \Delta w = \lambda w \quad \text{in} \quad \Omega \ ,$$

(1.19) $$w = 0 \quad \text{on} \quad \partial\Omega \ .$$

It is well known that $\lambda_1(\rho)$ is simple, i.e., it has multiplicity one, and the corresponding eigenfunctions do not change sign in Ω. For $\rho \equiv 0$ we will denote by φ_1 the eigenfunction of λ_1 such that $\varphi_1 > 0$ and $\|\varphi_1\|_{L^\infty(\Omega)} = 1$ (normalization condition). Hence we have

(1.20) $$- \Delta\varphi_1 = \lambda_1\varphi_1 \ , \quad \varphi_1 > 0 \quad \text{in} \quad \Omega \ ,$$

(1.21) $$\varphi_1 = 0 \qquad \text{on} \quad \partial\Omega \ .$$

A preliminary result concerning the existence of solutions of (1.11),(1.12) can be proved by using a comparison argument.

<u>Lemma 1.4</u>: Problem (1.11),(1.12) has no nontrivial solution for $\lambda \le \lambda_1$.

<u>Proof</u>: Suppose u is a nontrivial solution. We can define the

function

$$(1.22) \qquad h(x) = \begin{cases} \dfrac{f(u(x))}{u(x)} & \text{for } u(x) \neq 0 , \\ 0 & \text{for } u(x) = 0 . \end{cases}$$

(After, this function will be denoted by $[\frac{f(u)}{u}]$). It is not very difficult to show that h is continuous, and (1.14) yields $h \geq 0$, $h \not\equiv 0$. Hence

$$(1.23) \qquad -\Delta u + hu = \lambda u \quad \text{in } \Omega ,$$

$$(1.24) \qquad u = 0 \quad \text{on } \partial\Omega ,$$

and by the above mentioned comparison results

$$\lambda \geq \lambda_1(h) > \lambda_1(0) = \lambda_1 .$$

<u>Corollary 1.1</u>: If u is a nontrivial positive solution of (1.11), (1.12), then $u > 0$ in Ω.

<u>Proof</u>: If $u \geq 0$ is a solution of our problem, then u is a solution of (1.23),(1.24) with $h \geq 0$ continuous. By the Maximum Principle, $u \equiv 0$ or $u > 0$ in Ω.

Our next aim is to use Theorem 1.1 to prove the existence of positive solutions of (1.11),(1.12). (A completely similar argument could be used to study negative solutions, but this is not interesting for our purposes).

Thus, we need a subsolution u_0 and a supersolution u^0 satisfying $0 < u_0 \leq u^0$. For the supersolution we may choose $u^0 \equiv M$, with $M > 0$ sufficiently large. Indeed, M is a supersolution if $f(M) - \lambda M \geq 0$ and (1.15) implies that such a M exists. More precisely, if we put $k(x) = \frac{f(x)}{x}$, $x \neq 0$, and $g = k^{-1}$ (which exists and is strictly increasing by (1.14)),we may choose as M

$$(1.25) \qquad M_\lambda = g(\lambda) .$$

This will be useful below.

As a subsolution we pick $u_o \equiv c\,\varphi_1$, with φ_1 given by (1.20), (1.21) and $c > 0$ small enough. Indeed,

$$- \Delta u_o + f(u_o) - \lambda u_o = c\,\varphi_1 \,[\,\lambda_1 - \lambda + \frac{f(c\varphi_1)}{c\,\varphi_1}\,]$$

and for $\lambda > \lambda_1$ fixed it is easy to see that for $c > 0$ sufficiently small, u_o is a subsolution.

<u>Theorem 1.2</u>: Problem (1.11),(1.12) has at least a nontrivial positive solution for any $\lambda > \lambda_1$.

<u>Proof</u>: The assertion follows immediately from Theorem 1.1 and the obvious fact that $u_o < u^o$ for c (resp. M) small (resp. large) enough.

The following theorem states that for every $\lambda > \lambda_1$ our problem has a unique nontrivial positive solution, this result will follow essentially from (1.14).

<u>Theorem 1.3</u>: If $\lambda > \lambda_1$, then problem (1.11) (1.12) has a unique nontrivial positive solution.

<u>Proof 1 ([37],[108],[61])</u>: Take $\lambda > \lambda_1$ fixed. Existence has been already proved in Theorem 1.2. Suppose that u and v are nontrivial positive solutions of (1.11) (1.12). Hence, by Corollary 1.1, $u > 0$ and $v > 0$ in Ω .

We consider first the case, e.g., $u \le v$. We multiply (1.11) by v and integrate over Ω by using Green's Formula and (1.12). Doing the same thing with (1.11) (1.12) for v , and u , we get

$$\int_\Omega - \Delta u \,.\, v + f(u)v = \lambda \int_\Omega uv = \int_\Omega \nabla u \,.\, \nabla v + f(u)v =$$

$$\int_\Omega - \Delta v \,.\, u + f(v)u = \int_\Omega \nabla u \,.\, \nabla v + f(v)u$$

and this yields

$$\int_\Omega f(u)v - f(v)u = 0 \;.$$

If we put

$$A = \{x \in \Omega \mid 0 < u(x) < v(x)\}$$

$$B = \{x \in \Omega \mid 0 < u(x) = v(x)\}$$

it is clear that the integral over B is 0 , and hence

$$(1.26) \qquad \int_\Omega f(u)v - f(v)u = \int_A f(u)v - f(v)u = \int_A uv\left(\frac{f(u)}{u} - \frac{f(v)}{v}\right) = 0$$

and by (1.14) and the continuity of u and v, $u \equiv v$.

Suppose now that u and v are not ordered, i.e., $u \not\leq v$ and $v \not\leq u$. In this case, as it was shown earlier, it is possible to find a supersolution $M > 0$ such that $u < M$, $v < M$. By applying Theorem 1.1 with this $M > 0$ and the same subsolution as before, we prove the existence of a maximal solution $\bar{u}$ such that $u \leq \bar{u}$, $v \leq \bar{u}$. But now, by the first part of the proof $u \equiv v \equiv \bar{u}$.

<u>Proof 2 ([20])</u>: It follows easily from the Maximum Principle that if $u \geq 0$ is a solution, then $u \leq M_\lambda$, with M_λ given by (1.25). Hence there is a <u>maximum</u> solution, which we will call u_λ, and it is the unique positive solution. Indeed, if $u \geq 0$ is a solution, then $u \leq u_\lambda$, $u \not\equiv u_\lambda$ in Ω, and by (1.14), $\frac{f(u)}{u} \leq \frac{f(u_\lambda)}{u_\lambda}$. Reasoning as in Lemma 1.4 we get

$$\lambda = \lambda_1\left(\left[\frac{f(u)}{u}\right]\right) < \lambda_1\left(\left[\frac{f(u_\lambda)}{u_\lambda}\right]\right) = \lambda ,$$

a contradiction.

<u>Proof 3 ([37])</u>: By assumption (1.14), it follows that the function $\lambda u - f(u)$ is concave, and then the uniqueness is an easy consequence of the results in [37].

<u>Remark 1.6</u>: The first part of Proof 1 works equally if u and v are two ordered positive solutions (i.e., $u \leq v$ or $v \geq u$) and f satisfies, instead of (1.14), which means strict convexity, a strict concavity assumption, namely $\frac{f(u)}{u}$ strictly decreasing for $u > 0$ and

strictly increasing for $u < 0$. Then $u \equiv v$. But the second part, involving sub and supersolutions, does not work any more if they are not avalaible. We will return to this problem in Section 3.1.

Remark 1.7: The same method can be applied if (1.15) is replaced by the assumption that f is asymptotically linear (cf. (3.6) below). The main difference in this case is that a different trick should be employed to find supersolutions (cf. [61]).

We have obtained some results concerning existence and uniqueness of nontrivial positive solutions of (1.11) (1.12) depending on the values of the real parameter λ. It is possible to obtain, in exactly the same way, the corresponding results for nontrivial negative solutions.

The next propositions provide some complementary results concerning qualitative properties of the "branch" of positive solutions. These matters will be considered again in these notes (cf. in particular Section 3.1). In fact, the following proposition says that if $u(\lambda)$ is the only nontrivial positive solution for the value λ of the parameter, then $u(\lambda)$ is an increasing continuous "branch".

Proposition 1.1: The mapping $\lambda \longrightarrow u(\lambda)$ from $[\lambda_1, +\infty)$ into $C(\bar{\Omega})$ is continuous. Moreover, if $\lambda \leq \mu$, then $u(\lambda) \leq u(\mu)$ (pointwise).

Proof: It is an easy exercise by using sub and supersolutions and the continuity of T. Cf., e.g., [61],[20].

Remark 1.8: This is by no means the best smoothness result for the branch $u(\lambda)$. We will show below (Section 3.1) that $u(\lambda)$ is actually C^2 as a consequence of (1.13) (1.14). On the other hand, the monotonicity result can be made more precise by using the Maximum Principle.

Remark 1.9: It is also possible to prove that λ_1 is the only bifurcation point for positive solutions, cf. [10],[12],[61].

Another useful result is contained in the following proposition. Its proof, which uses sub and supersolutions and comparison arguments, can be found in [20].

Proposition 1.2: If $x \in \Omega$, $u(\lambda)(x) \xrightarrow[\lambda \to +\infty]{} +\infty$. Moreover

$$g(\lambda - \lambda_1) \leq \|u(\lambda)\|_{L^\infty(\Omega)} \leq g(\lambda) \quad . \tag{1.27}$$

Finally, a very simple comparison argument (cf. [20]) shows that, in the interval $\lambda_1 < \lambda < \lambda_2$, the three solutions we know, namely the trivial solution, the unique positive solution and the unique negative solution, are the only ones. A more complicated proof of this result was given before in [13] by using topological methods, cf. Section 2.2. Cf. also [76].

Proposition 1.3: If $\lambda_1 < \lambda < \lambda_2$, then problem (1.11),(1.12) with f satisfying (1.13)-(1.15) has exactly 3 solutions.

Proof: If u is a nontrivial solution, then u satisfies

$$\begin{aligned} - \Delta u + [\frac{f(u)}{u}]\, u &= \lambda u && \text{in } \Omega \\ u &= 0 && \text{on } \partial\Omega \, , \end{aligned}$$

Hence $\lambda = \lambda_i([\frac{f(u)}{u}])$ for some $i \geq 1$. But $\frac{f(u)}{u} > 0$, and this gives

$$\lambda_2 > \lambda = \lambda_i([\frac{f(u)}{u}]) > \lambda_i(0) = \lambda_i \quad .$$

Hence $i = 1$, and $u > 0$ or $u < 0$.

The results in this section can be summarized in the following diagram.

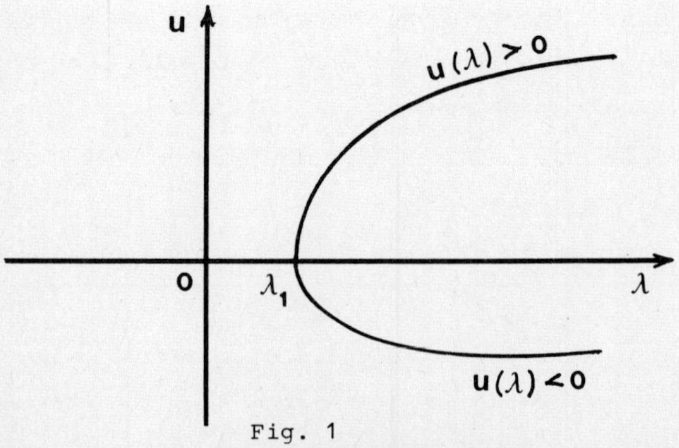

Fig. 1

1.3. THE METHOD OF SUB AND SUPERSOLUTIONS FOR SYSTEMS: APPLICATIONS.

We study in this Section the existence of solutions for some reaction-diffusion systems. For the sake of simplicity we only treat the case of systems with two equations with homogeneous Dirichlet boundary conditions. Let the system

$$(1.28) \qquad -\Delta u = f(u,v) \quad \text{in } \Omega ,$$

$$(1.29) \qquad -\Delta v = g(u,v) \quad \text{in } \Omega ,$$

$$(1.30) \qquad u = v = 0 \quad \text{on } \partial\Omega ,$$

where Ω is a smooth bounded domain in $\mathbb{R}^N$ and the nonlinear terms f and g satisfy the smoothness assumption

$$(1.31) \qquad f,\ g \in C^1 .$$

Now it seems natural, when trying to extend the method of sub and supersolutions to systems, to add some monotonicity assumptions on the growth of f and g as functions of u and v.

The first (apparently) reasonable extension is to assume that f and g are both, separately, increasing in the variables u and v. In this case, if we adopt an (apparently) reasonable definition of sub and supersolutions for systems, the method extends easily to the system (1.28)-(1.30). More precisely, if f and g satisfy (1.31) and are separately increasing in u and v, and if we define a subsolution (u_o,v_o) and a supersolution (u^o,v^o) of (1.28)-(1.30) as pair of smooth functions satisfying

$$(1.32) \qquad -\Delta u_o - f(u_o,v_o) \le 0 \le -\Delta u^o - f(u^o,v^o) \quad \text{in } \Omega ,$$

$$(1.33) \qquad -\Delta v_o - g(u_o,v_o) \le 0 \le -\Delta v^o - g(u^o,v^o) \quad \text{in } \Omega ,$$

$$(1.34) \qquad u_o \le 0 \le u^o , \quad v_o \le 0 \le v^o \quad \text{on } \partial\Omega ,$$

then, if $u_o \le u^o$, $v_o \le v^o$, the corresponding version of Theorem 1.1

is true and its proof can be carried out in a completely similar way, as was already pointed out in [100].

Moreover, it is always possible to make f (resp. g) increasing in u (resp. in v) by considering (cf. Remark 1.1) the equivalent problem

$$(1.35)\qquad -\Delta u + Mu = f(u,v) + Mu \quad \text{in } \Omega ,$$

$$(1.36)\qquad -\Delta v + Mv = g(u,v) + Mv \quad \text{in } \Omega ,$$

$$(1.37)\qquad u = v = 0 \quad \text{on } \partial\Omega ,$$

where $M > 0$ is such that the right-hand side in (1.35) (resp. in (1.36)) is increasing in u (resp. in v).

However, if f is not increasing in v, or if g is not increasing in u, then the nonlinear operator T defined (as in the proof of Theorem 1.1) by means of the system (1.35)-(1.37) is no more increasing, separately, in u and v, and the method of monotone iterations does not work. Here is the reason for the relevance of those monotonicity assumptions: if f (resp. g) is increasing in v (resp. in u) or, more generally, if each nonlinear term is increasing in the "off-diagonal" variables, the system is called <u>quasi-monotone</u>. There is a large literature on this subject.

Another possibility is to use, when monotonicity properties of f and g allow it, different tricks (changes of variable, particular iterative schemes, etc.) to obtain monotone sequences converging to maximal and minimal solutions. Cf., e.g., [78],[104],[86].

On the other side, if f and g do not satisfy suitable monotonicity assumptions, with the above definition of sub and supersolutions (1.32)-(1.34) the natural extension of Theorem 1.1 is false, as shows the following counterexample: the system we consider is

$$-\Delta u = u - v + 1 \quad \text{in } \Omega ,$$

$$-\Delta v = u \quad \text{in } \Omega ,$$

$$u = v = 0 \quad \text{on } \partial\Omega .$$

It is easy to see that, for this system, $(u_o,v_o) \equiv (0,0)$ is a subsolution and $(u^o, v^o) \equiv (0,1)$ is a supersolution. Hence, if the theorem is still valid in this case, there is a solution $(\bar{u}, \bar{v})$ with $\bar{u} \equiv 0$ and $0 \le \bar{v} \le 1$. But if $\bar{u} \equiv 0$, then $\bar{v} \equiv 0$ on the one side and $\bar{v} \equiv 1$ on the other, a contradiction

This counterexample seems to indicate that the above definition of sub and supersolution (1.32)-(1.34) is too weak, at least without additional monotonicity assumptions. An idea about how to modify the definition of sub and supersolution can be obtained by looking, from a somewhat different point of view, at the proof of Theorem 1.1. Indeed, by the order-preserving properties of the operator T (Lemmas 1.1-1.3) it follows immediately that the interval

$$K = [u_o, u^o] = \{z \mid u_o(x) \le z(x) \le u^o(x) \quad \text{in} \quad \Omega\}$$

is invariant by T, i.e., $T(K) \subset K$, and this suggests to use, e.g., Schauder's fixed point theorem if the corresponding assumptions on T and K are satisfied. Of course, this result would be weaker than Theorem 1.1, but its application to systems could be interesting.

We remark that, even if the result is weaker, it supposes some additional restrictions on the function space we employ. The proof of Theorem 1.1 was carried out in $C^{1,\alpha}(\bar{\Omega})$, but many other function spaces, as it was pointed out, were equally well-adapted, e.g., $C^{\alpha}(\bar{\Omega})$, $C(\bar{\Omega})$ or $L^2(\Omega)$. To apply Schauder's fixed point theorem we need a compact operator T and a closed, bounded, convex subset K. There is no problem with T. But the interval K is not bounded in $C^{1,\alpha}(\bar{\Omega})$ (and even in $C^{\alpha}(\bar{\Omega})$, cf., e.g. [10, p. 629], while it is in $C(\bar{\Omega})$ or $L^2(\Omega)$. (The use of these spaces would imply, as usual, some regularity argument to show that the weak solutions we obtained are actually classical solutions). The "reason" for these difficulties is that the norms involve the derivatives in some cases, while the order involves only the functions. But, with a suitable choice of the function space and the definition of sub and supersolution, it is possible to prove an existence theorem for systems generalizing Theorem 1.1 without monotonicity assumptions on f and g.

<u>Definition</u>: The pair $(u_o,v_o)-(u^o,v^o)$ is called a <u>sub-supersolution</u> (or a coupled sub-supersolution) of system (1.28)-(1.30) if $u_o, u^o, v_o, v^o \in H^2(\Omega) \cap L^\infty(\Omega)$, $u_o \le u^o$, $v_o \le v^o$ and

$$(1.38) \qquad -\Delta u_o - f(u_o,v) \le 0 \le -\Delta u^o - f(u^o,v) \qquad \forall\, v \in [v_o,v^o]$$

$$(1.39) \qquad -\Delta v_o - g(u,v_o) \le 0 \le -\Delta v^o - g(u,v^o) \qquad \forall\, u \in [u_o,u^o]$$

$$(1.40) \qquad u_o \le 0 \le u^o , \qquad v_o \le 0 \le v^o ,$$

where we use the notation

$$[K,\ell] = \{z \in L^2(\Omega) \mid K(x) \le z(x) \le \ell(x) \quad \text{a.e. on } \Omega\} .$$

<u>Remark 1.10</u>: It is clear that there is a coupling between the components of a sub-supersolution, contrarily to the case of definition (1.32)-(1.34). However, both coincide when f and g are, separately, increasing in u and v; if not, the last is much more stringent than the preceding one (cf. Remark 1.12) for an illuminating geometrical interpretation).

<u>Remark 1.11</u>: The idea of this definition comes from some classical work by Müller for systems of ordinary differential equations. Some different versions of it have appeared in works concerning both parabolic and elliptic reaction-diffusion systems, often in a simplified form due to the monotonicity properties of the nonlinear terms. Cf. [79],[62], [86].

<u>Remark. 1.12</u>: A very important particular instance of sub-supersolution is the case in which every component is a constant function: this is equivalent to say that the sub-supersolution determines an <u>invariant rectangle</u>, where the vector field (f,g) is always pointed to the interior of the rectangle along its boundary. The same geometrical interpretation in the case of definition (1.32)-(1.34) says that only the vectors corresponding to the four vertices are pointed to the interior. This notion of invariant rectangle is very important for the study

of reaction-diffusion systems, cf. [35],[106].

Theorem 1.4 ([62]): Suppose that $(u_o,v_o)-(u^o,v^o)$ is a sub-super-solution of the system (1.28)-(1.30) with f and g satisfying (1.31). Then there exists at least a (classical) solution (u,v) of the system such that $u_o \le u \le u^o$, $v_o \le v \le v^o$.

Proof: Let $E = [L^2(\Omega)]^2$ and $K = [u_o,u^o] \times [v_o,v^o]$. Then K is closed, bounded, convex subset of E. We define a nonlinear operator

$$T : K \longrightarrow E \quad ,$$

$$(\bar{u},\bar{v}) \longrightarrow (w,z) = T(\bar{u},\bar{v}) \quad ,$$

where (w,z) is the (unique) solution of the linear system

$$(1.41) \qquad -\Delta w + Mw = f(\bar{u},\bar{v}) + M\bar{u} \quad \text{in } \Omega \quad ,$$

$$(1.42) \qquad -\Delta z + Mz = g(\bar{u},\bar{v}) + M\bar{v} \quad \text{in } \Omega \quad ,$$

$$(1.43) \qquad w = z = 0 \quad \text{on } \partial\Omega \quad ,$$

with $M > 0$ such that the right-hand side in (1.41) (resp. (1.42)) is increasing in $\bar{u}$ (resp. in $\bar{v}$).

First, T is well-defined. Indeed, by the classical linear theory, as $f(\bar{u},\bar{v}) + M\bar{u} \in L^\infty(\Omega)$, there is a unique $w \in W^{2,p}(\Omega)$, for any $p \ge 1$, solution of (1.41) (1.43). The same for z. By the classical regularity theory, it follows in a quite straightforward way that T is compact.

It remains to prove that $T(K) \subset K$, i.e., that $u_o \le w \le u^o$ and $v_o \le z \le v^o$. We only show that $u_o \le w$, the other three inequalities can be proved in a completely analogous way. By (1.38) with $v = \bar{v}$ and (1.41)

$$0 \ge -\Delta(u_o-w) - f(u_o,\bar{v}) + f(\bar{u},\bar{v}) + M\bar{u} - Mw$$

$$= -\Delta(u_o-w) + f(\bar{u},\bar{v}) - f(u_o,\bar{v}) + M(\bar{u}-u_o) + M(u_o-w) \; .$$

We multiply this inequality by $(u_o-w)^+$ (we employ the usual notation $u^+ = \max(u,0)$, $u = u^+ - u^-$) and integrate over Ω using Green's Formula, and (1.40),(1.43)

$$0 \geq \int_\Omega -\Delta(u_o-w)(u_o-w)^+ + \int_\Omega [f(\bar{u},\bar{v}) - f(u_o,\bar{v}) + M(\bar{u}-u_o)](u_o-w)^+$$

$$+ M\int_\Omega (u_o-w)(u_o-w)^+ \geq \int_\Omega |\nabla(u_o-w)^+|^2 - \int_{\partial\Omega} (u_o-w)^+ \frac{\partial(u_o-w)}{\partial n}$$

$$+ M\int_\Omega |(u_o-w)^+|^2 = \int_\Omega |\nabla(u_o-w)^+|^2 + M\int_\Omega |(u_o-w)^+|^2 ,$$

and hence $(u_o-w)^+ = 0$ taking into account that the second integral is positive by monotonicity and the boundary integral is zero. (We have used a classical result by Stampacchia, namely that if $u \in H^1(\Omega)$, then $u^+, u^- \in H^1(\Omega)$ and $\int_\Omega \nabla u^+ . \nabla u^- = 0$). By Schauder's fixed point theorem, there is at least a fixed point of T, which is a weak solution of the system.

Finally, we show that (u,v) is a classical solution if and only if it is a fixed point of T. It is obvious that any classical solution is a fixed point of T. Conversely, if (u,v) is a fixed point, then $u, v \in W^{2,p}(\Omega)$ for any $1 \leq p < \infty$, and, by Morrey's Lemma, belong to $C^{1,\delta}(\bar{\Omega})$ for any $0 < \delta < 1$. Hence $f(u,v), g(u,v) \in C^1(\bar{\Omega})$ and by Schauder's Theory, $u, v \in C^{2,\delta}(\bar{\Omega})$ for any $0 < \delta < 1$ and (u,v) is a classical solution.

<u>Remark 1.13</u>: The existence result in Theorem 1.4 can be extended so as to cover the case where f and g are not locally Lipschitz, and even maximal monotone graphs. For these generalizations, which are motivated by free boundary problems arising, e.g., in chemical reactions and mathematical biology (cf. <u>Comments</u>) cf. [53],[45].

<u>Remark 1.14</u>: Concerning uniqueness, it is not difficult to prove by a direct argument, that if f and g are globally Lipschitz and the corresponding Lipschitz constants are "small", then the solution is unique. It is also possible, by using nonlinear semigroups [9] or sub and supersolutions [62],to prove simultaneously that the solution

is unique and globally asymptotically stable.

Finally, we give an application of Theorem 1.4. Consider the system

$$- \Delta u + f(u)\, g(v) = 0 \quad \text{in } \Omega ,$$
$$- \Delta v - h(u)\, k(v) = 0 \quad \text{in } \Omega ,$$
$$u = v = 1 ,$$

where f, g, h, k are C^1. Assume that they satisfy

(1.44) $f(0) \leq 0 , \quad f(1) > 0 ,$

(1.45) $g(0) \geq 0 , \quad g(v) \geq 0 \quad \text{if} \quad v \geq 0 ,$

(1.46) $h(u) \geq 0 \quad \text{if} \quad 0 \leq u \leq 1 ,$

(1.47) There is $M > 0$ such that $0 \leq k(v) \leq M$ for $v \geq 0$.

Then it is easily seen that $(u_o,v_o) - (u^o,u^o)$, where $u_o \equiv v_o \equiv 0$, $u^o \equiv 1$ and v^o is the unique solution of the linear problem

$$- \Delta w = M \max_{0 \leq u \leq 1} h(u) \quad \text{in } \Omega ,$$
$$w = 1 \quad \text{on } \partial\Omega ,$$

is a sub- supersolution and then there is at least one solution (u,v) of our problem such that $0 \leq u \leq 1$, $0 \leq v \leq v^o$. In particular, (1.44)-(1.47) are satisfied for $f(u) = \alpha u^p$, $h(u) = \alpha\beta u^p$, $g(v) = k(v) = e^{\frac{\gamma(v-1)}{v}}$, $\alpha, \beta, \gamma > 0$, $p \geq 1$, and for $f(u) = h(u) = u^p$,

$$g(v) = \alpha_1 e^{\frac{\gamma_1(v-1)}{v}} + \alpha_2 e^{\frac{\gamma_2(v-1)}{v}} ,$$

$$k(v) = \alpha_1\beta_1 e^{\frac{\gamma_1(v-1)}{v}} + \alpha_2\beta_2 e^{\frac{\gamma_2(v-1)}{v}} , \quad \alpha_1,\alpha_2,\beta_1,\beta_2,\gamma_1,\gamma_2 > 0 , \quad p \geq 1.$$

These systems arise in chemical reactions ([9],[60],[62]). They can be extended by using, e.g., the existence results in [80], as was pointed out in [53].

II. TOPOLOGICAL METHODS: THE LERAY-SCHAUDER DEGREE. GLOBAL BIFURCATION THEOREMS AND APPLICATIONS.

This chapter is devoted to topological methods, in particular the Leray-Schauder degree, and some of their applications to the nonlinear elliptic problems we are studying. The topological degree in infinite dimensional spaces was introduced in a very famous paper by Leray-Schauder [77] in 1934, generalizing the topological degree in finite dimension due to Brouwer.

The Leray-Schauder degree is one of the more general and powerful methods to prove existence theorems for nonlinear equations. Contrarily to the method of sub and supersolutions in the first chapter, comparison arguments do not play an important role here; in particular this implies that applications to nonlinear partial differential equations are not limited to second order equations where the Maximum Principle holds. An important example of this situation are (fourth order) Von Karman's equations in nonlinear elasticity, where comparison results are not available.

The topological degree is a very useful tool to prove existence results for nonlinear problems; from this point of view the main difficulty is usually to obtain a priori estimates on the solutions in order to be able to apply the theory. But the degree can also be used to "count" the number of solutions, obtaining lower bounds, or even giving the exact number of solutions (cf. Theorem 2.2 below). Moreover, the degree is the main tool to prove the so-called global bifurcation theorems, which are very useful, together with a priori estimates and maybe some additional information, to prove existence results.

Section 2.1 gives an informal account of the definition of topological degree, followed by a list of its main properties. These properties are used in Section 2.2 to compute the exact number of solutions of the nonlinear eigenvalue problem in Section 1.2. Section 2.3 includes (without proofs) a global bifurcation theorem by Rabinowitz and some applications, in particular to the same problem in Section 1.2. The next section gives similar results for the case without bifurcation.

The second part of this Chapter is especially concerned with positive solutions. Section 2.5 is devoted to global bifurcation theorems for positive solutions, Section 2.6 treats the fixed point index, and finally Section 2.7 contains an application to reaction-diffusion systems.

2.1. THE TOPOLOGICAL DEGREE. MAIN PROPERTIES.

In this Section we intend to give only a very general and informal idea of the definition of the topological degree, first in finite dimension, and then in the infinite dimensional case. For a systematic exposition, including full proofs, cf. [97],[83],[51],[22].

Let Ω be a bounded open subset of $\mathbb{R}^N$ $(N \geq 1)$, and let $\varphi : \Omega \longrightarrow \mathbb{R}^N$, $\varphi \in C^1(\Omega) \cap C(\bar{\Omega})$. We define $S = \{x \in \Omega \mid J_\varphi(x) = 0\}$, where $J_\varphi(x)$ denotes the Jacobian of φ at the point $x \in \Omega$, which is called the set of <u>singular points</u> of φ. If $b \notin \varphi(\partial\Omega) \cup \varphi(S)$, it is easy to see that $\varphi^{-1}(b)$ is a discrete compact subset, and hence finite. Consequently the number

$$d(\varphi,\Omega,b) = \sum_{x \in \varphi^{-1}(b)} \text{sg } J_\varphi(x) ,$$

where sg denotes the sign, is well- defined since the sum in the right-hand side is finite. The integer $d(\varphi,\Omega,b)$ is called the <u>degree</u> of the map φ relative to the point b with respect to Ω. Defined in this way, the degree can be considered as a kind of "algebraic" counting of the number of solutions of the equation $\varphi(x) = b$ in Ω.

We would like to eliminate two restrictions in this definition. The first one, $b \notin \varphi(S)$, is easy to overcome by using Sard's theorem ($\varphi(S)$ has Lebesgue measure 0). The second is more important: we want to define the degree for every continuous map φ. Then φ is approximated (uniformly) by C^1 functions, and it is necessary to check that the degree is independent of the approximation.

Before to generalize the degree to the infinite dimensional case, it is convenient to remark that it is impossible to define there a topological degree for every continuous map with the same properties as in finite dimension. Indeed, if not, Brouwer's fixed point theorem (which can be proved by using degree theory, cf., e.g., [97]) would be still valid in infinite dimension, and well-known counterexamples show that this is not the case. Schauder's fixed point theorem (or theorems), which, in some sense, play its role, require some compactness assumption: if $T(K) \subset K$ and K is a closed, bounded, convex subset, T should be compact; if T is only continuous, then K should be compact (and convex).

It seems then necessary, in order to have an extension to infinite dimension of the topological degree, to work in some subclass of the class of all continuous maps involving some compactness condition: the set of all mappings of the form $I - T$, with T compact, usually called compact perturbations of the identity or "compact vector fields" in the terminology of [73].

Let E be a real Banach space, and let $T : \Omega \longrightarrow E$, where Ω is a bounded open subset of E, with T compact. Then it is possible to define the degree of the map $\Phi = I - T$ by using, roughly speaking, the fact that compact operators can be approximated by operators with finite dimensional range. Then, if $b \notin \Phi(\partial\Omega)$, it is possible to define an integer $d(\Phi,\Omega,b)$, the Leray-Schauder degree of $\Phi = I - T$ relative to the point b with respect to Ω, with the following properties:

1. Continuity with respect to T: There exists a neighborhood V of T in $Q(\Omega,E)$ (space of compact mappings from Ω into E with the norm $\|T\| = \sup_{x\in\bar{\Omega}} \|Tx\|$) such that for any $S \in V$, $b \notin (I-S)(\partial\Omega)$ and

$$d(I - S,\Omega,b) = d(\Phi,\Omega,b) .$$

2. Homotopy invariance: Let $H \in C(\bar{\Omega} \times [0,1],E)$, where $H(u,t) = u - S(u,t)$, with $S \in Q(\bar{\Omega} \times [0,1],E)$. If $b \notin H(\partial\Omega \times [0,1])$,

then for any $t \in [0,1]$

$$d(H(.,t),\Omega,b) = \text{const.} \quad .$$

3. The degree is constant on connected components of $E - \Phi(\partial\Omega)$.

4. Additivity: Let $\Omega = \Omega_1 \cup \Omega_2$, with Ω_1 and Ω_2 disjoint bounded open subsets of E. If $b \notin \Phi(\partial\Omega_1) \cup \Phi(\partial\Omega_2)$, then

$$d(\Phi,\Omega,b) = d(\Phi,\Omega_1,b) + d(\Phi,\Omega_2,b) \ .$$

5. $d(\Phi,\Omega,b) = \begin{cases} 1 & \text{if } b \in \Omega \\ 0 & \text{if } b \notin \bar{\Omega} \end{cases}$.

6. If $b \notin \Phi(\bar{\Omega})$, then $d(\Phi,\Omega,b) = 0$.

7. Existence: If $d(\Phi,\Omega,b) \neq 0$, then there exists a $x_o \in \Omega$ such that $\Phi(x_o) = b$.

8. Excision: If $K \subset \Omega$ is closed and $b \notin \Phi(K)$, then

$$d(\Phi,\Omega,b) = d(\Phi,\Omega - K,b) \ .$$

We are now in the position to introduce another topological tool, easily derived from the degree, which will also be useful in the following, the index or local degree.

Let $\Phi = I - T$ as above and let u_o be an isolated solution of the equation $\Phi(u) = b$, i.e., a solution such that it is the unique solution in some neighborhood of u_o. By the excision property

$$d(\Phi,B_r(u_o),b) = d(\Phi,B_{r_o}(u_o),b)$$

for every $0 < r < r_o$, r_o sufficiently small, and the index of the isolated solution u_o is defined by

$$i(\Phi,u_o,b) = d(\Phi,B_r(u_o),b) \quad .$$

The following results are very useful, since they allow to calculate the index, and then the degree, in some situations, and this is very convenient.

Theorem 2.1: Let $L : E \longrightarrow E$ be a compact linear operator and let $\Phi = I - L$. If 1 is not a characteristic value of L, then

$$i(\Phi,0,0) = (-1)^{\beta} ,$$

where β is the sum of the (algebraic) multiplicities of the characteristic values of L lying in $(0,1)$.

Remark that $i(\Phi,0,0)$ is well-defined, since $\Phi(u) = 0$ implies $u = 0$.

Lemma 2.1: Let $T : \Omega \longrightarrow E$, $0 \in \Omega$, and let T compact and Fréchet-differentiable at 0. Then $T'(0)$ is a compact linear operator.

Corollary 2.1: Let $\Phi = I - T$, $T \in Q(\Omega,E)$, where Ω is a neighborhood of 0. Suppose that $To = 0$ and that T is Fréchet-differentiable at 0. If 1 is not a characteristic value of $T'(0)$, then 0 is an isolated solution of $\Phi(u) = 0$ and $i(\Phi,0,0) = (-1)^{\beta}$, where β is the sum of the (algebraic) multiplicities of the characteristic values of $T'(0)$ lying in $(0,1)$.

Corollary 2.2: Suppose that T satisfies the assumptions in Corollary 2.1 and let $\Phi_{\lambda}(u) = u - \lambda Tu$. Then $i(\Phi_{\lambda},0,0)$ is well-defined if λ is not a characteristic value of $T'(0)$ and it changes by $(-1)^{m_j}$ when λ crosses a characteristic value μ_j of (algebraic) multiplicity m_j .

2.2. AN APPLICATION: THE EXACT NUMBER OF SOLUTIONS OF A NONLINEAR EIGENVALUE PROBLEM.

In this Section the topological tools of the preceding will be applied to the study of the exact number of solutions of the nonlinear eigenvalue problem (1.11) (1.12) already considered in Section 1.2. This problem was settled by Proposition 1.3, whose proof was indeed very simple. We include this alternative (longer) proof to illustrate how to use the degree in such situations. Moreover, this method can be also applied in more general problems where comparison results are not avalaible (cf. [13]).

We consider again the problem

$$(2.1) \qquad - \Delta u + f(u) = \lambda u \quad \text{in} \quad \Omega \ ,$$

$$(2.2) \qquad u = 0 \quad \text{on} \quad \partial\Omega \ ,$$

where Ω is a smooth bounded domain in $\mathbb{R}^N$, λ is a real parameter, and $f : \mathbb{R} \longrightarrow \mathbb{R}$ satisfies the assumptions

(2.3) f is C^2, increasing, and $f(0) = f'(0) = 0$,

(2.4) $\frac{f(u)}{u}$ is strictly increasing for $u > 0$ and strictly decreasing for $u < 0$,

$$(2.5) \qquad \lim_{|u| \to +\infty} \frac{f(u)}{u} = + \infty \ .$$

We restate Proposition 1.3.

Theorem 2.2: If $\lambda_1 < \lambda < \lambda_2$, then problem (2.1) (2.2), with f satisfying (2.3)-(2.5), has exactly 3 solutions.

Roughly speaking, the proof can be described as follows: for λ fixed, $\lambda_1 < \lambda < \lambda_2$, problem (2.1) (2.2) can be reformulated, once again, in an equivalent form, namely to find the fixed points of an associated compact nonlinear operator T in $C(\bar{\Omega})$ (or in $L^2(\Omega)$, cf. [76]). Suppose that we claim that the solutions of $u - Tu = 0$ have the properties:

i) Every solution is isolated,

ii) If $u = 0$, $i(I - T, 0, 0) = -1$,

iii) If $u \neq 0$, $i(I - T, u, 0) = +1$,

iv) There is an a priori estimate for solutions of (2.1),(2.2) and $d(I - T, B_R(0), 0) = +1$ for $R > 0$ sufficiently large.

Now the theorem follows readily from these claims by a simple counting.

First, we reformulate our problem as a fixed point equation $u - Tu = 0$. The problem can be written as

$$(2.6) \qquad \begin{aligned} -\Delta u &= \lambda u - f(u) && \text{in } \Omega , \\ u &= 0 && \text{on } \partial\Omega . \end{aligned}$$

We are led to define the Nemitskii operator associated with the right-hand side in (2.6), $F : C(\bar{\Omega}) \longrightarrow C(\bar{\Omega})$, where for $u \in C(\bar{\Omega})$, Fu is defined by $Fu(x) = \lambda u(x) - f(u(x))$. It is well-known (cf., e.g., [10]) that F is continuous and bounded (i.e., A bounded implies $F(A)$ bounded). If we write $L = (-\Delta)^{-1}$, then (2.6) is equivalent to

$$(2.7) \qquad u = LF(u) = Tu ,$$

where $T = LF : C(\bar{\Omega}) \longrightarrow C(\bar{\Omega})$ is compact since L is compact.

The following series of lemmas proves the claims i)-iv) and this gives the proof of Theorem 2.2.

<u>Lemma 2.2</u>: There exists $R > 0$ such that if u is a solution of (2.1),(2.2) with $\lambda_1 < \lambda < \lambda_2$, then $\|u\|_{L^\infty(\Omega)} < R$.

The proof, which is not very difficult, can be found in [98],[76], [50]. For the case of positive solutions, cf. Section 3.1 below.

<u>Lemma 2.3 ([76])</u>: If $u \neq 0$ is a solution of (2.1),(2.2), then, for any $w \in H_o^1(\Omega)$, $w \not\equiv 0$

$$0 < \int_\Omega (|\nabla w|^2 - \lambda w^2 + f'(u)w^2)\, dx .$$

<u>Remark 2.1</u>: Lemma 2.3 means that, for $\lambda_1 < \lambda < \lambda_2$, the linearized operator along a nontrivial solution is invertible or, in other words, that these solutions are not degenerated.

<u>Lemma 2.4</u>: There exists $R > 0$ such that $u - Tu = 0$ implies $\|u\|_{L^\infty(\Omega)} < R$. Moreover, $d(I - T, B_R(0), 0) = 1$.

<u>Proof</u>: The first part is simply Lemma 2.2. For the second, let $t \in [0,1]$ and let u satisfying $u - t\,Tu = 0$. Consequently

$$- \Delta u = t(\lambda u - f(u)) \quad \text{in } \Omega ,$$
$$u = 0 \quad \text{on } \partial\Omega ,$$

and hence it is clear that the estimate for $t = 1$ (Lemma 2.2) yields the estimate for any $t \in [0,1]$. By the homotopy invariance for the degree

$$d(I - tT, B_R(0), 0) = d(I - T, B_R(0), 0) = d(I, B_R(0), 0) = 1.$$

Lemma 2.5: If $u_o - Tu_o = 0$, $u_o \neq 0$, then u_o is isolated and $i(I - T, u_o, 0) = 1$.

Proof: By Corollary 2.1 it is sufficient to show that $\gamma \geq 1$ and γ eigenvalue of $T'(u_o)$, i.e., $T'(u_o)w = \gamma w$, imply $w \equiv 0$. Indeed, for $\gamma = 1$ this means that 1 is not a characteristic value, and for $\gamma > 1$ that no characteristic value is in $(0,1)$. Hence $i(I - T, u_o, 0) = 1$.

Let us check this. The equation $T'(u_o)w = LF'(u_o)w = \gamma w$ is equivalent to

$$- \gamma\Delta w = \lambda w - f'(u_o)w \quad \text{in } \Omega$$
$$w = 0 \quad \text{on } \partial\Omega .$$

Multiplying by w and integrating over Ω we obtain $(\gamma \geq 1)$ by Lemma 2.3

$$\int_\Omega \gamma|\nabla w|^2 - \lambda w^2 + f'(u_o)w^2 = 0 \geq \int_\Omega |\nabla w|^2 - \lambda w^2 + f'(u_o)w^2 > 0$$

if $w \not\equiv 0$, a contradiction. Hence $w \equiv 0$.

Lemma 2.6: The solution $u = 0$ is isolated and $i(I - T, 0, 0) = -1$.

Proof: In a similar way, since $f'(0) = 0$, the equation $w = T'(0)w$ is equivalent to $-\Delta w = \lambda w$ in Ω, $w = 0$ on $\partial\Omega$. Since $\lambda_1 < \lambda < \lambda_2$, $w \equiv 0$. By Corollary 2.1 $u = 0$ is isolated and $i(I - T, 0, 0) = (-1)^\beta$, where β is the sum of the multiplicities of the characteristic values in $(0,1)$. Then $\gamma w = T'(0)w$, $w \not\equiv 0$, $\gamma \geq 1$, is equivalent to $\frac{\lambda}{\gamma} = \lambda_j$

and $\gamma = \frac{\lambda}{\lambda_j} \geq 1$. Hence $j = 1$, $\beta = 1$ and finally $i(I-T,0,0) = -1$.

2.3. GLOBAL BIFURCATION THEOREMS.

Let E be a real Banach space and let $\Phi : \mathbb{R} \times E \longrightarrow E$ smooth. Assume that $\phi(\lambda,0) = 0$ for every $\lambda \in \mathbb{R}$. The problem $\Phi(\lambda,u) = 0$ has, for all values of the real parameter λ, the solution $u = 0$, which is usually called the trivial solution. We only consider problems of the form

(2.8) $$\Phi(\lambda,u) = u - G(\lambda,u) = 0 ,$$

where

(2.9) $$G(\lambda,u) = \lambda Lu + H(\lambda,u) ,$$

with L and H satisfying

(2.10) L is a compact linear operator,

(2.11) H is compact and $H(\lambda,u) = o(\|u\|)$ at $u = 0$, uniformly for λ bounded, i.e., $\lim_{u\to 0} \frac{H(\lambda,u)}{\|u\|} = 0$, uniformly for λ on bounded intervals.

In this situation it is well-known that to ensure the point λ_o (or $(\lambda_o,0)$) is a bifurcation point of (2.9), a necessary condition is that λ_o is a characteristic value of L (cf., e.g., [73]). Very simple counter-examples show that this condition is not sufficient. There are some rather general sufficient conditions: one of them is that λ_o is a characteristic value of L of odd (algebraic) multiplicity (cf. [73]). But this result has a purely local character: all that it means is that there is a (maybe "small") neighborhood of $(\lambda_o,0)$ in $\mathbb{R} \times E$ such that every "sub-neighborhood" contains a solution (λ,u) of (2.9) with $u \neq 0$. The following theorem, due to Rabinowitz, shows that the fact that λ_o has odd multiplicity has much more deep implications, concerning in particular the global structure of the solution set of (2.9).

Theorem 2.3 ([91],[92]): Assume (2.10) (2.11). If λ_0 is a characteristic value of odd (algebraic) multiplicity of L, then there exists a connected component C of S (the closure in $\mathbb{R} \times E$ of the set of nontrivial solutions of (2.9)) such that C contains $(\lambda_0, 0)$ and either C is unbounded either contains $(\hat{\lambda}_0, 0)$, where $\hat{\lambda}_0$ is a characteristic value of L, $\hat{\lambda}_0 \neq \lambda_0$.

Roughly speaking, this is the plan of the proof: if the theorem is false, then it follows by some topological arguments the existence of a one-parameter family of mappings such that, on the one side they have the same degree by homotopy invariance and, on the other, satisfy the assumptions of Corollary 2.2. This implies that the index should change from 1 to -1 (or from -1 to 1) when crossing an odd characteristic value, a contradiction.

It is very easy to see that both cases in Theorem 2.3 actually occur. If $H \equiv 0$, the problem is reduced to $u = \lambda Lu$, and the connected components are simply the corresponding eigenspaces. As an example of the second ([97]) we can take $E = \mathbb{R}^2$ and

$$u = \lambda Lu + LHu \qquad u = \begin{pmatrix} u_1 \\ u_2 \end{pmatrix}$$

with

$$L = \begin{pmatrix} 1 & 0 \\ 0 & 1/2 \end{pmatrix} \qquad H(u) = \begin{pmatrix} -u_2^3 \\ u_1^3 \end{pmatrix} .$$

L has the characteristic values 1 and 2, both simple, which are bifurcation points. A very simple calculation shows that the set of nontrivial solutions is a closed loop in the interval $1 < \lambda < 2$.

Remark 2.2: Similar results can be proved for bifurcation at infinity, cf. [95].

The main difficulty in order to apply Theorem 2.3 is to know which of the two above possibilities actually occurs. In two very interesting cases (a class of nonlinear Sturm-Liouville problems and a class of quasi-linear elliptic problems) it is possible to show (cf. [91], [92],[97]) that the continua of solutions bifurcating from simple eigenvalues (as it is the case for all the eigenvalues for Sturm-Liouville problems and the first for some elliptic problems) are unbounded. The proof exploits largely some qualitative properties of the solutions. More precisely, in the case of Sturm-Liouville problems, the <u>nodal</u> properties of the eigenfunctions: if φ_K is an eigenfunction for the eigenvalue λ_K, then φ_K has exactly $(K-1)$ simple zeroes in the corresponding interval. The same properties imply that continua arising from different eigenvalues cannot intersect. In the case of quasi-linear elliptic problems having a linearized eigenvalue problem as, e.g., (1.16) (1.17) such arguments only apply to the first eigenvalue, which is simple and has a positive eigenfunction.

Global bifurcation theorems can be used to obtain existence results for (2.9). However, we point out that, even if it is possible to prove that the continuum of solutions given by Rabinowitz's theorem is unbounded, the information on the solution set is not very precise. Indeed, all we know, in principle, about C is that C is a closed connected subset. Thus, it is very convenient to get as more additional information as possible on solutions.

A decisive step in this direction is to have <u>a priori</u> estimates for the solutions of (2.9). If there is a continuous function $\varepsilon : \mathbb{R} \longrightarrow \mathbb{R}^+$ such that $\Phi(\lambda,u) = 0$ implies $||u|| \leq \varepsilon(\lambda)$ and C is unbounded, then C can "go to infinity" "to the right" and/or "to the left", but not "vertically" (cf. Fig. 2).

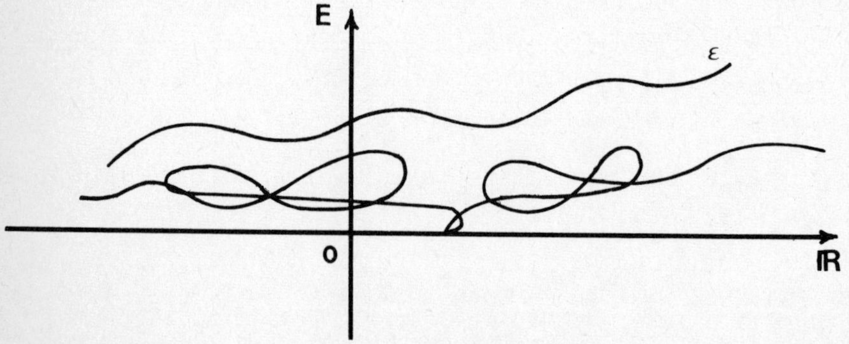

Fig. 2

Moreover, if some more information on solutions is available, then the structure of the solution set can be made more precise. Cf. Sections 1.2 and 3.1.

Let us apply this result to our problem (2.1) (2.2) with assumptions (2.3)-(2.5). The problem can be written as

$$u = \lambda Lu - L(f(u))$$

where $L = (-\Delta)^{-1}$. Since $f(0) = f'(0) = 0$ it is not difficult to check that for any "reasonable" norm (e.g. $C(\bar{\Omega})$, $C^{\alpha}(\bar{\Omega})$, etc.) $\frac{\|Lf(u)\|}{\|u\|} \xrightarrow[u \to 0]{} 0$, and the linearized problem $u = \lambda Lu$ has the simple eigenvalue λ_1. Now it is possible to apply Theorem 2.3 to λ_1 and to prove that the component C is unbounded. On the other hand, recall that we have some more information on solutions of (2.1) (2.2) at our disposal: for $\lambda \leq \lambda_1$ there is only the trivial solution (Lemma 1.4) and there are <u>a priori</u> estimates (Lemma 2.2): both imply that solutions should "go to infinity on the right". But anyway, even using these complementary results, the method of sub and supersolutions in Section 1.2 and the continuation method of Section 3.1 give much more information.

These considerations concerning our test problem are still valid in a rather general setting. When applicable, the method of sub and supersolutions seems to be preferable to global bifurcation arguments in order to prove existence results for nonlinear problems of the form (2.8): besides to give a constructive proof involving monotone iterations, they allow to get more precise information on the structure of the solution set. But if sub and supersolutions are not available, then topological methods, and in particular global bifurcation theorems, can be used to prove existence results.

2.4. EXISTENCE OF CONTINUA IN THE CASE WITHOUT BIFURCATION.

Let E be a real Banach space and let $T : \mathbb{R} \times E \longrightarrow E$ compact such that $T(0,u) = 0$ for every $u \in E$. Hence $(0,0)$ is the only solution of problem

$$(2.12) \qquad u = T(\lambda,u)$$

contained in $\{0\} \times E$. The following theorem is an example of the kind of results which can be proved in the non-bifurcation case.

Theorem 2.4 ([97]): With the above assumptions, if C is the connected component of the solution set of (2.12) containing $(0,0)$, then $C = C^+ \cup C^-$, where $C^\pm \subset \mathbb{R}^\pm \times E$, $C^\pm$ are unbounded and $C^+ \cap C^- = \{(0,0)\}$.

It is also possible to obtain similar results when $T(0,0) \neq 0$ if some a priori estimates are available. For example we can prove the following theorem (cf. [92]).

Theorem 2.5: Let $T : \mathbb{R} \times E \longrightarrow E$ compact and suppose there exists $M > 0$ such that $u = T(0,u)$ implies $\|u\| < M$. If $d(\Phi(0,.), B_M(0),0) \neq 0$, then S has two unbounded components in $\mathbb{R}^+ \times E$ and $\mathbb{R}^- \times E$ intersecting in $(0,\bar{u})$, where $\|\bar{u}\| < M$.

2.5. GLOBAL BIFURCATION THEOREMS FOR POSITIVE SOLUTIONS.

It is fairly well-known that positive solutions play a very important role in applications as, e.g., chemical reactions, combustion theory, population dynamics, etc., where some of the unknowns (concentrations, populations, etc.) are essentially positive.

Moreover, there are global bifurcation theorems for positive solutions excluding the second possibility in Theorem 2.3, ensuring in this way the existence of an unbounded component of (positive) solutions.

Suppose (E,P) is an ordered Banach space with positive cone P, P is generating (i.e., $E = P - P$) and let $T : \mathbb{R}^+ \times P \longrightarrow P$ compact such that $T(\lambda,0) = 0$ for every $\lambda \geq 0$ and $T(0,u) = 0$ for every $u \in P$. We study the equation

$$(2.13) \qquad T(\lambda,u) = \lambda Lu + H(\lambda,u)$$

where L is a positive compact linear operator and H is compact and such that $H(\lambda,u) = o(\|u\|)$ at $u = 0$ uniformly for λ bounded.

Theorem 2.6 ([10]): Suppose all the above assumptions are satisfied and suppose that L possesses exactly one positive characteristic value λ_o with a positive eigenvector. Then λ_o is the only bifurcation point for positive solutions for (2.13). Moreover, the set of positive solutions contains a connected component C which is unbounded and such that $C \cap (\mathbb{R}^+ \times \{0\}) = \{(\lambda_o,0)\}$.

Remark 2.3: There are similar results,concerning bifurcation at infinity (cf. [10]). In particular, it is possible to show that, under hypotheses similar to those in Theorem 2.6, there is a unique asymptotic bifurcation point for positive solutions and that the corresponding continuum is also unbounded.

There are results similar to those in Section 2.4 for positive solutions in the case without bifurcation. Let us give only one example.

Theorem 2.7 ([10]): Let (E,P) be an ordered Banach space with cone P and let $T : \mathbb{R}^+ \times P \longrightarrow P$ compact such that $T(0,u) = 0$ for every $u \in P$. Then the connected component of the solution set of $u = T(\lambda,u)$ containing $(0,0)$ is unbounded.

It is clear that Theorem 2.7 becomes trivial if $T(\lambda,0) = 0$ for every λ: in this case the solution set contains the "line" of trivial solutions $\mathbb{R}^+ \times \{0\}$. It is interesting in this setting to study the existence of positive solutions bifurcating from the trivial solution.

An interesting nonlinear problem, where some of the above methods can be tested is the equation

$$(2.14) \qquad \begin{aligned} -\Delta u &= \lambda e^u && \text{in } \Omega , \\ u &= 0 && \text{on } \partial\Omega , \end{aligned}$$

arising in the study of some simplified models for combustion (cf., e.g. [17]). This problem has been considered by several authors. In particular, Guelfand [59] proved that in the special case of the sphere $(N = 3)$ there are infinitely many solutions for a certain value $\bar{\lambda}$ of the parameter λ, and Joseph-Lundgren [67] provide a complete and interesting study of radial solutions by using ordinary differential equations methods.

Let $L = (-\Delta)^{-1}$ and let F be the Nemitskii operator (in $C(\bar{\Omega})$) associated to e^u, i.e., $Fu(x) = e^{u(x)}$. It is clear that (2.14) is equivalent to $u = \lambda LF(u)$, or to $u = T(\lambda,u)$, where $T(\lambda,u) = \lambda LF(u)$ is compact, and that $T(0,u) = 0$ for every u. On the other hand, it follows immediately from the Maximum Principle that if u is a solution of (2.14) for $\lambda \geq 0$, then $u > 0$ in Ω. Otherwise stated, since every solution is positive, the problem can be formulated in the above framework: in particular, by Theorem 2.7 it follows the existence of an unbounded continuum of positive solutions of (2.14) containing (0,0). This information, which is not as precise as we desire, can be completed by some additional considerations: e.g., there is a $\bar{\lambda} > 0$ such that (2.14) has no solution for $\lambda > \bar{\lambda}$ (cf. Lemma 3.1 and [44]).

2.6. THE FIXED POINT INDEX.

It was already remarked that the Leray-Schauder degree is one of the more powerful tools to prove existence results for nonlinear equations, and that it can be useful to compute the exact number of solutions (cf. Theorem 2.2) or at least a lower bound. However, in some situations, e.g., in problems arising in physical applications, the main interest is concentrated almost exclusively on positive solutions, and it would be nice to have at our disposal some analogous tools in order to work in a (relatively) open subset of a cone of positive functions. But if this cone has empty interior (as, e.g., in $L^2(\Omega)$), then the topological degree cannot be directly applied, since the corresponding open subset (the interior of the cone) is empty. Anyway, this difficulty can be overcomed by exploiting the fact that a cone is a retract in a Banach space: this allows us to define a fixed point index for compact operators defined on the cone.

First, we recall some definitions and theorems from general topology. Let X be a topological space and let $A \subset X$. Then A is called a <u>retract</u> of X if there exists a continuous map $r : X \longrightarrow A$ called <u>retraction</u> such that $r|_A = I_A$. It is easily seen that every retract is a closed subset.

Following a theorem by Dugundji, every closed convex subset of a Banach space E is a retract of E. This allows us to define a fixed point index (cf. [10],[85] for more details).

Let X be a retract of the Banach space E. Let U be a bounded open subset of X and let $f : \bar{U} \longrightarrow X$ compact which has no fixed points on ∂U. Then there exists an integer $i(f,U,X)$ with the properties:

1. <u>Normalization</u>: For every constant map $f : \bar{U} \longrightarrow U$, $i(f,U,X) = 1$.

2. <u>Additivity</u>: For every pair of disjoint bounded open subsets U_1 and U_2 of U such that f has no fixed points on $\bar{U} - (U_1 \cup U_2)$

$$i(f,U,X) = i(f,U_1,X) + i(f,U_2,X)$$

where $i(f,U_i,X) = i(f|_{\bar{U}_i}, U_i, X)$, $i = 1,2$.

3. <u>Homotopy invariance</u>: For every compact interval $I \subset \mathbb{R}$ and every compact map $H : I \times \bar{U} \longrightarrow X$ such that $H(\lambda,x) \neq x$ for $(\lambda,x) \in I \times \partial U$,

$$i(H(\lambda,.), U, X) = \text{const.}$$

for every $\lambda \in I$.

4. <u>Permanence</u>: If Y is a retract of X and $f(\bar{U}) \subset Y$, then

$$i(f,U,X) = i(f,U \cap Y, Y)$$

where $i(f,U \cap Y, Y) = i(f|_{\overline{U \cap Y}}, U \cap Y, Y)$.

5. <u>Excision</u>: For $V \subset U$, V bounded and open, such that f has no fixed point in $\bar{U} - V$,

$$i(f,U,X) = i(f,V,X).$$

6. <u>Existence</u>: If $i(f,U,X) \neq 0$, then f has at least a fixed point in U.

This integer is called the <u>fixed point index</u>. In the case of an ordered Banach space (E,P), the cone P is, following a theorem by Dugundji, a retract of E, and then the corresponding fixed point index is well- defined for a bounded open subset U of P and a compact map $f : \bar{U} \longrightarrow P$ with no fixed points on ∂U. In the following the fixed point index will be denoted by $i(f,U)$, omitting the reference to the cone P.

Remark 2.4: If U is an open subset of the space E, then $i(f,U,E) = d(I-f,U,0)$. This shows that the fixed point index is a rather natural generalization of the degree. Cf. [10].

2.7. AN APPLICATION: POSITIVE SOLUTIONS FOR A REACTION-DIFFUSION SYSTEM.

This Section is devoted to an application of the fixed point index to the existence of positive solutions of a nonlinear system arising in applications, more precisely in the study of chemical reactions (cf. [63]). A variant of the preceding global bifurcation theorems is needed for our problem, namely the following.

Lemma 2.7 ([63]): Let (F,P) be an ordered Banach space with cone P and let $f : \mathbb{R}^+ \times P \longrightarrow P$ compact. Suppose that there is a constant $M > 0$ such that $w = f(0,w)$ implies $\|w\| < M$ and that $i(f(0,.), B_M(0) \cap P) \neq 0$. Then there exist a continuum of solutions of $x = f(\lambda,x)$, which is unbounded in $\mathbb{R}^+ \times P$ and such that C contains $(0,\bar{w})$ where $\bar{w} \in P$ and $\|\bar{w}\| < M$.

We shall apply this abstract result to our example. Consider the system

$$(2.15) \qquad -a_1 \Delta u + \frac{\lambda u}{u+1} - Auv = 0 \quad \text{in } \Omega ,$$

$$(2.16) \qquad -a_2 \Delta v + Auv = 1 \quad \text{in } \Omega ,$$

$$(2.17) \qquad u = v = 0 \quad \text{on } \partial\Omega ,$$

where Ω is a smooth bounded domain in $\mathbb{R}^N$, $a_1, a_2, A > 0$ are real numbers and λ is a real parameter. Let $f(u) = \frac{u}{u+1}$, it is clear that for $u \geq 0$, $0 \leq f(u) \leq u$. We fix a_1, a_2 and A and take λ as a parameter.

We consider the function space $E = [C^{1,\alpha}(\bar{\Omega})]^2$ with the positive cone $K = \{(w,z) \in E \mid w,z \geq 0 \text{ in } \Omega\}$. We define a map

$$G : \mathbb{R}^+ \times K \longrightarrow K$$

$$(\lambda,u,v) \longrightarrow (w,z) = G(\lambda,u,v)$$

in the following way: for each $(\lambda,u,v) \in \mathbb{R}^+ \times K$, (w,z) is the (unique) solution of the system

$$(2.18) \qquad - a_1 \Delta w + \lambda w = Auv + \lambda(u-f(u)) \quad \text{in } \Omega ,$$

$$(2.19) \qquad - a_2 \Delta z + Auz = 1 \quad \text{in } \Omega ,$$

$$(2.20) \qquad w = z = 0 \quad \text{on } \partial\Omega .$$

Indeed, for $u,v \geq 0$, the system is decoupled in two linear equations. Moreover, by the Maximum Principle, $w,z > 0$ in Ω (recall that $f(u) \leq u$ if $u \geq 0$). The operator G is compact, and the proof is rather straightforward.

<u>Lemma 2.8</u>: There is a constant $c > 0$ which is independent of λ, such that if (u,v) is a solution of (2.15)-(2.17) with $\lambda \geq 0$ and $u,v \geq 0$, then

$$(2.21) \qquad ||u||_{L^\infty(\Omega)} \leq c , \qquad ||v||_{L^\infty(\Omega)} \leq c .$$

<u>Proof</u>: Let g be the solution of the problem

$$(2.22) \qquad - \Delta g = -1 \quad \text{in } \Omega ,$$

$$(2.23) \qquad g = 0 \quad \text{on } \partial\Omega .$$

By adding equations (2.15)-(2.17) and (2.22),(2.23) we obtain

$$- \Delta(a_1 u + a_2 v + g) = - \lambda f(u) \le 0 \quad \text{in} \quad \Omega \ ,$$

$$a_1 v + a_2 v + g = 0 \quad \text{on} \quad \partial\Omega \ .$$

Hence, by the Maximum Principle, $a_1 u + a_2 v + g \le 0$ in Ω, and then $0 \le a_1 u + a_2 v \le -g$, which gives the result.

Lemma 2.9: There is a constant $M > 0$ such that if $(u,v) \in K$ is a solution of (2.15)-(2.17) for $\lambda = 0$, then

$$(2.24) \qquad \|u\|_{1,\alpha} \le \frac{M}{2} \ , \quad \|v\|_{1,\alpha} \le \frac{M}{2} \ .$$

Proof: It follows easily from (2.21), the L^p estimates in [1] and Morrey's Lemma.

Theorem 2.8: There exists an unbounded component of positive solutions of (2.15)-(2.17) such that its projection on the real axis is all $\mathbb{R}^+$. In particular, for any $\lambda \ge 0$ there is at least a positive solution of (2.15)-(2.17).

Proof: We apply Lemma 2.7 with $F = E$, $P = K$, and $f = G$. G is compact and $G(0,u,v) = (u,v)$ implies (2.24) (Lemma 2.9). We claim that $i(G(0,.), B_M(0) \cap K) \ne 0$. Indeed, recall that $G(0,.)$ is given by (2.18)-(2.20) with $\lambda = 0$ and define the homotopy $(w,z) = H(t,u,v)$ as the solution of the (linear) system

$$- a_1 \Delta w = tAuv \quad \text{in} \quad \Omega \ ,$$

$$- a_2 \Delta z + tAuz = 1 \quad \text{in} \quad \Omega \ ,$$

$$w = z = 0 \quad \text{on} \quad \partial\Omega \ .$$

Since $H(1,.) = G(0,.)$ and, moreover, (2.24) implies $H(t,u,v) \ne (u,v)$ for any $t \in [0,1]$ and any $(u,v) \in \partial(B_M(0) \cap K)$, the homotopy invariance of the index yields

$$i(G(0,.), B_M(0) \cap K) = i(H(1,.), B_M(0) \cap K) = i(H(0,.), B_M(0) \cap K) = 1$$

since $H(0,.)$ is a constant map, by the normalization property.

Finally, we remark that a priori bounds for any $\lambda > 0$ can be easily obtained.

Remark 2.5: We point out that in the example (2.15)-(2.17), the discontinuity of the function $\frac{u}{u+1}$ at $u = -1$ raises serious problems concerning the associated Nemitskii operator. This is no more the case if we are restricted to $u \geq 0$. On the other hand, the fact that we only consider positive solutions makes easier to obtain a priori bounds (cf. the proof of Lemma 2.8). Similar arguments, but working on the whole space and not on a cone, were used in [82] for the Brusselator.

III. CONTINUATION METHODS. EXISTENCE OF A SECOND SOLUTIONS: VARIATIONAL METHODS.

Continuation and variational methods are presented in this third chapter. Section 3.1 treats our test problem (the nonlinear eigenvalue problem (1.11),(1.12)) by following a different approach involving the use of a local inversion theorem by Crandall-Rabinowitz and a continuation argument which is an application of the Implicit Function Theorem. Once again, a priori estimates play an important role. This method is particularly well-suited for the study of bifurcation from simple eigenvalues, as it is the case here. We also study some variants of problem (1.11), (1.12), which led to some open problems.

Moreover, the same kind of methods are still useful in problems without bifurcation. In Section 3.2 we consider a class of problems including equation (2.14). We only sketch the arguments, sending to [44] for a careful study of the "bending points". Finally, Section 3.3 introduces a new method, namely Lyusternik-Schnirelman critical point theory, which, maybe combined with some other tools, is very useful to prove existence and multiplicity results.

3.1. LOCAL INVERSION THEOREMS AND CONTINUATION: THE BIFURCATION CASE.

We will consider here, for the third time, the nonlinear eigenvalue problem (1.11), (1.12) which was treated by using sub and supersolutions in Section 1.2 and by using degree theory and global bifurcation theorems in Section 2.2 and 2.3, respectively. We remark that, even if the idea of using local (and global) inversion theorems to prove existence results is not very recent (Hadamard, Caccioppoli, etc.), we follow here the approach by Rabinowitz [90], cf. also [42].

Consider once again the problem

$$- \Delta u + f(u) = \lambda u \quad \text{in} \quad \Omega \quad , \tag{3.1}$$

$$(3.2) \qquad u = 0 \quad \text{on} \quad \partial\Omega$$

where Ω is a smooth bounded domain in $\mathbb{R}^N$, λ is a real parameter, and $f : \mathbb{R} \longrightarrow \mathbb{R}$ satisfies

(3.3) f is C^2, increasing, and $f(0) = f'(0) = 0$,

(3.4) $\frac{f(u)}{u}$ is strictly increasing for $u > 0$ and strictly decreasing for $u < 0$,

$$(3.5) \qquad \lim_{|u| \to +\infty} \frac{f(u)}{u} = +\infty$$

or

$$(3.6) \qquad \lim_{|u| \to +\infty} \frac{f(u)}{u} = f'(\infty) < +\infty .$$

Lemma 3.1: Assume that f satisfies (3.3), (3.4) and (3.5) (resp. (3.6)). Then, if u is a nontrivial positive solution of (3.1), (3.2), $\lambda_1 < \lambda$ (resp. $\lambda_1 < \lambda < \lambda_1 + f'(\infty)$).

Proof: The first part is just Lemma 1.4. The second follows by a similar comparison argument.

The following theorem is the main tool for the results in this Section. Cf. also [42],[14],[4],[10].

Theorem 3.1 ([43]): Let X and Y be real Banach spaces, let $I \subset \mathbb{R}$ a bounded interval and let $F : I \times X \longrightarrow Y$, $F \in C^2$. Let $\lambda_o \in I$ and assume that F satisfies

i) $F(\lambda, 0) = 0$ for every $\lambda \in I$,

ii) $\dim \operatorname{Ker} (F_x(\lambda_o, 0)) = \operatorname{codim} R(F_x(\lambda_o, 0)) = 1$,

iii) $F_{\lambda x}(\lambda_o, 0)x_o \notin R(F_x(\lambda_o, 0))$, where x_o spans $\operatorname{Ker} F_x(\lambda_o, 0)$ (this will be denoted by $[x_o] = \operatorname{Ker} F_x(\lambda_o, 0)$).

Let Z be a complementary subspace of $[x_o]$ in X . Then there exist an interval J containing the origin and two C^1 functions $\lambda : J \longrightarrow \mathbb{R}$ and $\psi : J \longrightarrow Z$ such that $\lambda(0) = \lambda_o$, $\psi(0) = 0$ and

and $x(s) = sx_o + s\psi(s)$ implies $F(\lambda(s),x(s)) = 0$. Moreover, $F^{-1}(0)$ is uniquely formed (in a neighborhood of $(\lambda_o,0)$) by the curves $x = 0$ and $(\lambda(s),x(s))$, $s \in J$.

Theorem 3.2 ([61]): Suppose that f satisfies (3.3), (3.4), (3.5) (resp. (3.6)). Then, for any λ such that $\lambda_1 < \lambda$ (resp. $\lambda_1 < \lambda < \lambda_1 + f'(\infty)$) there exist a unique nontrivial positive solution $u(\lambda)$ of (3.1), (3.2). The mapping $\lambda \longrightarrow u(\lambda)$ from $[\lambda_1,+\infty)$ (resp. from $[\lambda_1,\lambda_1 + f'(\infty)))$ into $C_o^{2,\alpha}(\bar{\Omega})$ is C^2. Moreover, if f satisfies (3.6), then

$$\lim_{\lambda \uparrow \lambda_1 + f'(\infty)} \|u(\lambda)\|_\alpha = +\infty .$$

Proof: First, we recall that the uniqueness was already proved in Theorem 1.3.

In the first part of the proof, the local inversion Theorem 3.1 is applied to show the existence of a "small" branch of positive solutions bifurcating to the right from $\lambda = \lambda_1$. Moreover, this branch can be parametrized by λ.

We define the function space

$$C_o^{2,\alpha}(\bar{\Omega}) = \{z \in C^{2,\alpha}(\bar{\Omega}) \,|\, z = 0 \text{ on } \partial\Omega\}$$

and the mapping

$$F : \mathbb{R} \times C_o^{2,\alpha}(\bar{\Omega}) \longrightarrow C^\alpha(\bar{\Omega})$$

by

$$F(\lambda,u) = -\Delta u + f(u) - \lambda u .$$

It is clear that $F(\lambda,0) = 0$ for any λ and F is C^2 by (3.3). It is easily calculated that

$$F_u(\mu,v)w = -\Delta w + f'(u)w - \mu w$$

and

$$F_{\lambda u}(\mu,v)w = -w .$$

In particular

$$F_u(\lambda_1,0) = -\Delta w - \lambda_1 w$$

and hence

$$\mathrm{Ker}\ F_u(\lambda_1,0) = [\varphi_1] ,$$

with $\varphi_1 > 0$ on Ω (recall (1.20),(1.21)). It is easily seen (cf. [90]) that we can take

$$Z = R(F_u(\lambda_1,0)) = \{ z \in C^{\alpha}(\bar{\Omega}) \mid \int_{\Omega} z\, \varphi_1 = 0 \} .$$

Since $\varphi_1 > 0$ on Ω, $F_{\lambda u}(\lambda_1,0)\varphi_1 = -\varphi_1 \notin Z$ and we apply Theorem 3.1 with $X = C_o^{2,\alpha}(\bar{\Omega})$, $Y = C^{\alpha}(\bar{\Omega})$ and $\lambda_o = \lambda_1$. Hence there exist an interval $J \subset \mathbb{R}$ containing the origin and C^1 functions $\lambda : J \longrightarrow \mathbb{R}$, $\psi : J \longrightarrow Z$ such that $\lambda(0) = \lambda_1$, $\psi(0) = 0$, $u(s) = s\,\varphi_1 + s\,\psi(s)$ implies $F(\lambda(s),u(s)) = 0$ for $s \in J$; moreover, in a neighborhood of $(\lambda_1,0)$ all the nontrivial solutions are of this form.

We claim that these solutions are positive and that they can be parametrized by λ. If $\lambda(s) = \lambda_1 + \xi(s)$, where obviously $\xi(s) = \lambda(s) - \lambda_1$, then it is possible to see that, for $s > 0$ sufficiently small, $u(s) > 0$ on Ω, and by Lemma 3.1, $\xi(s) > 0$. Hence there exists a $s_o > 0$ such that the map $\lambda : (0,s_o) \longrightarrow \mathbb{R}$ is one-to-one. Indeed, if $\lambda(s) = \lambda(s')$, then by uniqueness $u(s) = u(s')$ and then $s = s'$ by the form of $u(s)$. This means that the curve of positive solutions can be parametrized by λ in a neighborhood on the right of λ_1.

In the second part we extend to the right this branch by using a continuation argument. Let (μ,v) be a solution such that $v > 0$. Then

$$(3.7) \qquad -\Delta v + \frac{f(v)}{v} v = \mu v \quad \text{in } \Omega ,$$
$$v = 0 \quad \text{on } \partial\Omega .$$

On the other hand, $F_u(\mu,v)$ is an isomorphism. Indeed, if not there exists a $w \not\equiv 0$ such that

$$(3.8) \qquad \begin{aligned} -\Delta w + f'(v)w &= \mu w \quad \text{in } \Omega \\ w &= 0 \quad \text{on } \partial\Omega . \end{aligned}$$

But, by (3.3),(3.4), $\frac{f(v)}{v} < f'(v)$ and hence $(v > 0)$ we obtain by usual comparison arguments

$$\mu \geq \lambda_1(f'(v)) > \lambda_1 \left(\frac{f(v)}{v} \right) = \mu ,$$

which is a contradiction. Hence $F_u(\mu,u)$ is an isomorphism and we apply the Implicit Function Theorem to obtain the existence of an $\varepsilon > 0$ and a C^2 mapping $\lambda \longrightarrow u(\lambda)$ defined for $|\lambda - \mu| < \varepsilon$ such that $F(\lambda,u(\lambda)) = 0$.

We have to prove that the solutions obtained prolongating in this way our initial branch are actually positive. By the continuity of $\lambda \longrightarrow u(\lambda)$ it follows that if $u(\lambda)$ is not in $K = \{w \in C_o^{2,\alpha}(\bar{\Omega}) \,|\, w > 0$ in $\Omega\}$ for all the values of λ, then there should be a $\bar{\lambda}$ minimal such that $\lambda_1 < \bar{\lambda}$ and $u(\bar{\lambda}) \in \partial K$, i.e., $u(\bar{\lambda})(a) = 0$ for some $a \in \Omega$. By Corollary 1.1, $u(\bar{\lambda}) \equiv 0$ and moreover $\|u(\lambda)\|_\alpha \longrightarrow 0$ when $\lambda \uparrow \bar{\lambda}$. This implies (cf. Theorem 2.6 and [10]) that $\bar{\lambda}$ is the only bifurcation point for positive solutions and $\bar{\lambda} = \lambda_1$, which is impossible.

Hence we have showed the existence of a branch of positive solutions for every λ in a maximal interval of the form (λ_1,λ^+), where $\lambda^+ \leq +\infty$.

We claim that (3.3), (3.4), (3.6) imply $\lambda^+ = \lambda_1 + f'(\infty)$. By Lemma 3.1, $\lambda^+ \leq \lambda_1 + f'(\infty)$. On the other hand

$$(3.9) \qquad \lim_{\lambda \uparrow \lambda^+} \|u(\lambda)\|_\alpha = +\infty .$$

Indeed, if not there are a constant c and a sequence λ_n such that $\lambda_n \uparrow \lambda^+$ and $\|u(\lambda_n)\|_\alpha \leq c$. A rather straightforward argument involving the compactness of the Green's operator (or solution operator) of (3.1),(3.2) shows that it is possible to apply the Implicit Function

Theorem in λ^+, contradicting the maximality of λ^+. By (3.9) and the results in [61] it follows that λ^+ is an asymptotic bifurcation point, and then $\lambda^+ = \lambda_1 + f'(\infty)$. Cf. [61],[12] for details.

If f satisfies (3.3), (3.4), (3.5), then it is clear that a necessary and sufficient condition for $\lambda^+ = +\infty$ is the existence of <u>a priori</u> estimates for solutions of (3.1), (3.2), i.e., the existence of a continuous function $\varphi : \mathbb{R}^+ \longrightarrow \mathbb{R}^+$ such that

$$\|u(\lambda)\|_\alpha \le \varphi(\lambda) \quad .$$

The existence of such a φ follows easily from Lemma 2.2 and the usual regularity theory. (Cf. also Section 1.2). However, as we only need <u>a priori</u> estimates for <u>positive</u> solutions, we use an alternative method. Suppose that $w > 0$ is a solution for the value λ of the parameter. Let $M = \max_{\bar{\Omega}} w = w(a) > 0$ for some $a \in \Omega$. Hence

$$-\Delta w(a) = \lambda w(a) - f(w(a)) \ge 0 \; ,$$

or, equivalently, $f(M) \le \lambda M$, and (with the notation of Section 1.2) then $M \le g(\lambda)$. This L^∞ estimate, together with the L^p and C^α regularity estimates, yields C^α (and even $C^{2,\alpha}$) estimates.

The proof of Theorem 3.2 gives an alternative approach to the study of positive solutions of our problem. (By the way, the same method can be used for the branch of negative solutions). In particular, we improve substantially the results concerning the smoothness of the branch $u(\lambda)$ as a function of λ: it was claimed in Proposition 1.1 that $u(\lambda)$ is continuous, but it is actually C^2 (and C^∞ if f is C^∞).

Now we study the same problem but with f strictly concave instead of strictly convex (assumption (3.4)), and try to apply the same method.

More precisely, we assume that $f : \mathbb{R} \longrightarrow \mathbb{R}$ satisfies the assumptions

(3.10) $\quad f$ is C^2, increasing, $f(0) = 0$, $f'(0) > 0$,

(3.11) $\quad \dfrac{f(u)}{u}$ is strictly decreasing for $u > 0$ and strictly increasing for $u < 0$,

$$(3.12) \qquad \lim_{|u|\to+\infty} \frac{f(u)}{u} = f'(\infty) \quad .$$

It is easily seen that in this case

$$(3.13) \qquad f'(\infty) < f'(0) \quad .$$

A first result concerning non-existence of positive solutions can be proved by using the same comparison arguments as in Lemma 1.4.

Lemma 3.2: Assume that f satisfies (3.10)-(3.12). Then, if u is a nontrivial solution of (3.1), (3.2), $\lambda_1 + f'(\infty) < \lambda < \lambda_1 + f'(0)$.

A first, and nontrivial difficulty arises in the study of the uniqueness of nontrivial positive solution, which is essential to apply this continuation method, since none of the proofs of Theorem 1.3 works. Indeed, now the function $\lambda u - f(u)$ is convex and sub and supersolutions are no more available. But (recall Remark 1.6), the first part of Proof 1 of Theorem 1.3 still works: if u and v are nontrivial positive solutions of (3.1),(3.2) such that, e.g., $u \le v$, then by (3.11) the same argument yields $u \equiv v$. On the other hand, the second part, involving sub and supersolutions, does not work. However, if it is possible to prove that for two nontrivial positive solutions u and v, $u \le v$ or $v \le u$, then we have uniqueness. This will be a consequence of the following additional assumption

$$(3.14) \qquad \lambda_1 + f'(0) < \lambda_2 + f'(\infty)$$

which can be written equivalently as $f'(0) - f'(\infty) < \lambda_2 - \lambda_1$. This assumption is exactly of the same kind of those in [14] or [7].

Lemma 3.3: Assume that f satisfies (3.10)-(3.12) and (3.14). If u and v are nontrivial positive solutions of (3.1),(3.2), then $u \le v$ or $v \le u$.

Proof: If u and v are solutions we have

$$-\Delta(u-v) + f(u) - f(v) = \lambda(u-v) \quad \text{in } \Omega \ ,$$

$$w = 0 \quad \text{on } \partial\Omega \ .$$

and if we put $w = u - v$, this can be rewritten as

$$(3.15) \qquad -\Delta w + aw = \lambda w \quad \text{in } \Omega, \qquad w = 0 \quad \text{on } \partial\Omega,$$

where a is defined as follows

$$a(x) = \begin{cases} \dfrac{f(u(x)) - f(v(x))}{u(x) - v(x)} & \text{if } u(x) \neq v(x), \\ f'(0) & \text{if } u(x) = v(x). \end{cases}$$

Then a is continuous and (3.11) and the definition of a imply $f'(\infty) \le a \le f'(0)$, $a \not\equiv f'(\infty)$.

We claim that if λ satisfies

$$\lambda_1 + f'(\infty) < \lambda \le \lambda_2 + f'(\infty),$$

then λ is the first eigenvalue of problem (3.15). Indeed, if not the usual comparison results imply

$$\lambda \ge \lambda_2(a) > \lambda_2(f'(\infty)) = \lambda_2 + f'(\infty),$$

a contradiction. Hence the corresponding eigenfunction $w = u - v$ does not change sign in Ω and, e.g., either $w > 0$, i.e., $u > v$, or either $w \equiv 0$, $u \equiv v$.

<u>Theorem 3.3 ([61])</u>: Suppose that f satisfies (3.10)-(3.12) and (3.14). Then, for any λ such that

$$(3.16) \qquad \lambda_1 + f'(\infty) < \lambda < \lambda_1 + f'(0)$$

there exists a unique nontrivial positive solution $u(\lambda)$ of (3.1),(3.2). The mapping $\lambda \longrightarrow u(\lambda)$ from $(\lambda_1 + f'(\infty), \lambda_1 + f'(0)]$ into $C_o^{2,\alpha}(\bar{\Omega})$ is C^2. Moreover

$$(3.17) \qquad \lim_{\lambda \downarrow \lambda_1 + f'(\infty)} \|u(\lambda)\|_\alpha = +\infty.$$

<u>Proof</u>: The uniqueness follows easily from Lemmas 3.2 and 3.3. The existence proof is quite similar to that in Theorem 3.2, so we only sketch it.

First, by a completely similar application of Theorem 3.1 we obtain the existence of a "small" branch of nontrivial positive solutions in a left-neighborhood of $\lambda_1 + f'(0)$. This branch can be parametrized by λ, and can be continued to the left by using again the Implicit Function Theorem; the proof that $F_u(\mu,v)$, where (μ,v) is a solution with $v > 0$, is an isomorphism is very similar, but this time $f'(v) < \frac{f(v)}{v}$ by (3.11). The proofs that the solutions on this continuated branch are positive, that there is a minimal $\bar{\lambda}$ for this continuation, and that $\bar{\lambda} = \lambda_1 + f'(\infty)$ are analogous.

The existence results in Theorems 3.2 and 3.3 can be summarized in the following diagrams (cf. Figures 3 to 5). We point out that $u(\lambda)$ is an increasing function of λ in the first case (cf. Proposition 1.1) but we do not know if the same is true, i.e., if $u(\lambda)$ is decreasing in λ, under the assumptions of Theorem 3.3, in Fig. 5.

Figure 3 corresponds to Theorem 3.2 with assumption (3.5) and Figure 4 to the same theorem with assumption (3.6). Figure 5 illustrates the situation for Theorem 3.3.

It remains the case of a function f satisfying the same hypotheses (3.10)-(3.12), but not the additional hypothesis (3.14), which play a decisive role in the uniqueness proof. Suppose that f satisfies (3.10)-(3.12) and

$$\lambda_2 + f'(\infty) < \lambda_1 + f'(0) .$$

In this situation, the first part of the argument is untouched, and we prove exactly as before the existence of a "small" branch of nontrivial positive solutions, parametrized by λ, in a left-neighborhood of $\lambda_1 + f'(0)$. Moreover, if (μ,v) is a solution with $v > 0$ and $\mu \le \lambda_2 + f'(\infty)$, then $F_u(\mu,v)$ is an isomorphism (the proof is the same) and the Implicit Function Theorem can be applied. In this way it can be proved, by using the fact $\lambda_1 + f'(\infty)$ is an asymptotic bifurcation point for positive solutions, that for any λ in the interval $(\lambda_1 + f'(\infty), \lambda_2 + f'(\infty)]$ there is a unique nontrivial positive solution. However, we do not know how to connect this branch with the "small" branch of positive solutions bifurcating from $\lambda_1 + f'(0)$, (Cf. Figure 6). On the other hand, it was proved in [61] by using the results in [3],[74] that there is <u>at least</u> a nontrivial positive solution for any λ in the interval $[\lambda_2 + f'(\infty), \lambda_1 + f'(0))$.

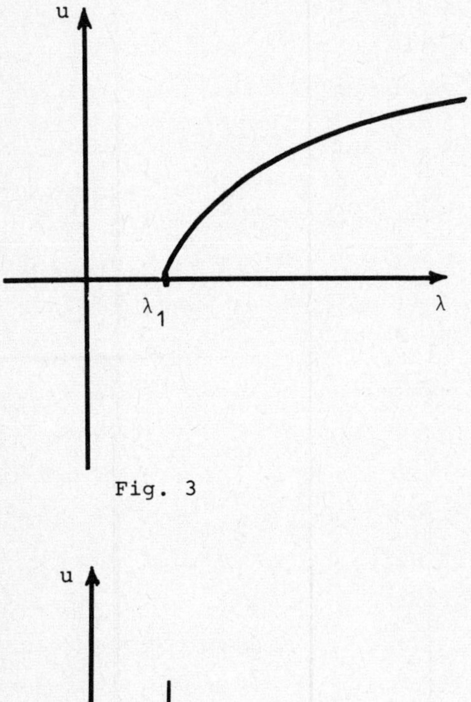

Fig. 3

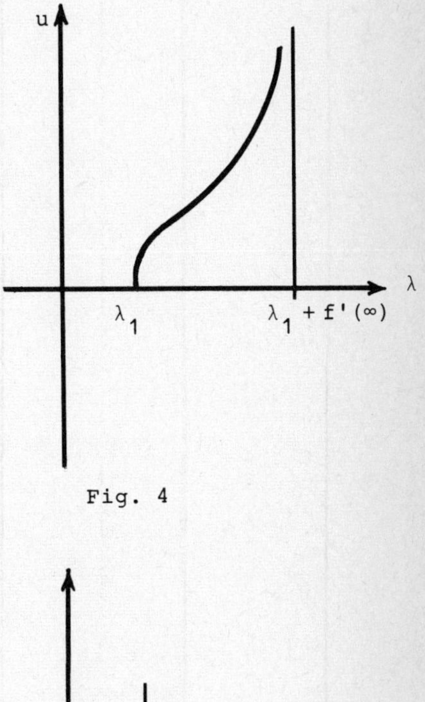

Fig. 4

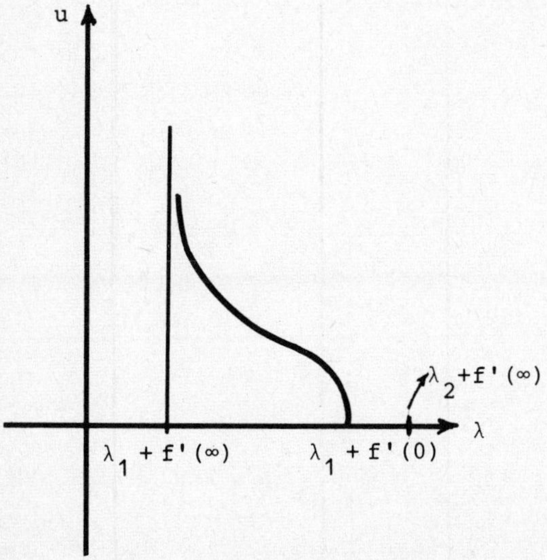

Fig. 5

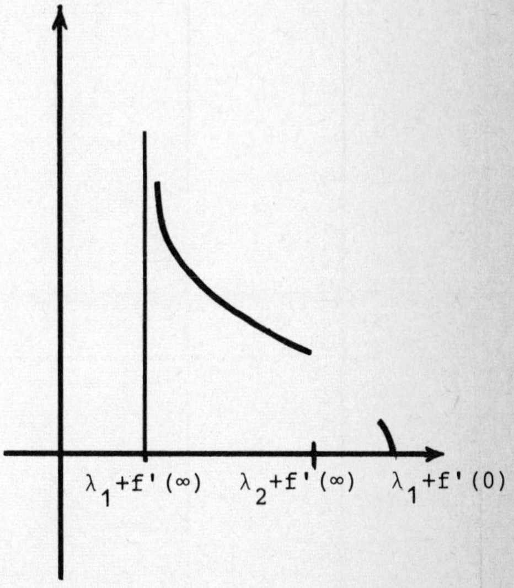

Fig. 6

3.2. LOCAL INVERSION THEOREMS AND CONTINUATION: THE CASE WITHOUT BIFURCATION.

We consider the nonlinear eigenvalue problem

(3.18) $$-\Delta u = \lambda f(x,u) \quad \text{in} \quad \Omega ,$$

(3.19) $$u = 0 \quad \text{on} \quad \partial\Omega ,$$

where the function $f : \bar{\Omega} \times \mathbb{R} \longrightarrow \mathbb{R}$ satisfies

(3.20) $$f \text{ is } C^2, \quad f(x,0) > 0 \quad \text{for any} \quad x \in \bar{\Omega} ,$$

(3.21) $$f_z(x,0) > 0 \quad \text{for any} \quad x \in \bar{\Omega} ,$$

(3.22) $$f_{zz}(x,z) > 0 \quad \text{for any} \quad x \in \bar{\Omega} \quad \text{and any} \quad z > 0 .$$

This kind of problems arise in several physical applications (cf. [59],[17]). An interesting particular case, namely $f(x,u) = e^u$, was considered in Section 2.5. But in some respects the case $f(x,u) = (1 + \gamma u)^\beta$, $\beta,\gamma > 0$, is more difficult to treat (cf. [67],[44]).

It was pointed out in Section 2.5 that for $\lambda > 0$ every solution u is strictly positive on Ω , and that there is an unbounded component of positive solutions containing (0,0). Moreover, by using sub and supersolutions and the obvious fact that 0 is a subsolution for (3.18), (3.19), it follows the existence of a minimal positive solution for some range of values of the parameter λ (cf. [10],[44]). But we are interested in getting some more information about solutions of (3.18),(3.19).

Let $\mu_1 > 0$ be the first eigenvalue of the linearized problem

$$- \Delta w = \mu f_z(x,0) w \quad \text{in} \quad \Omega ,$$

$$w = 0 \quad \text{on} \quad \partial\Omega .$$

By using comparison and continuation arguments it is not difficult to prove the following result (cf. [69],[44]).

<u>Theorem 3.4</u>: If u is a solution of (3.18),(3.19) with $\lambda > 0$,

then $\lambda \leq \mu_1$. Moreover, there is a number $0 < \bar{\lambda} \leq \mu_1$ which is maximal with respect to the existence of a curve $\lambda \longrightarrow u(\lambda)$ of positive solutions with the properties

i) the mapping $\lambda \longrightarrow u(\lambda)$ from $[0,\bar{\lambda})$ into $C_o^{2,\alpha}(\bar{\Omega})$ is continuous,

ii) the operator $v \longrightarrow -\Delta v - \lambda f_z(x,u(\lambda)(x))v$ is invertible as a map from $C_o^{2,\alpha}(\bar{\Omega})$ into $C^{\alpha}(\bar{\Omega})$ for $0 \leq \lambda < \bar{\lambda}$.

Finally, $u(\lambda)$ is increasing as a function of λ and, for fixed λ, $0 \leq \lambda < \bar{\lambda}$, is the minimal positive solution of (3.18),(3.19).

It is interesting to see the behaviour of the curve of solutions near the critical value $\bar{\lambda}$ of the parameter. The special cases e^u and $(1+\gamma u)^{\beta}$ were treated in [67] for spheres by using phase plane techniques and some of these results were extended to general bounded domains Ω in [44]. This is the main theorem in [44].

<u>Theorem 3.5</u>: Suppose there exists $M > 0$ such that $\|u(\lambda)\|_{L^{\infty}(\Omega)} \leq M$, where $u(\lambda)$ is the curve of minimal positive solutions given by Theorem 3.4. Then there exists $\bar{u} = \lim_{\lambda \uparrow \bar{\lambda}} u(\lambda)$ in the topology of $C_o^{2,\alpha}(\bar{\Omega})$, and there exists $\delta > 0$ such that the solutions of (3.18),(3.19) near $(\bar{\lambda},\bar{u})$ are on a curve $(\lambda(s),\tilde{u}(s))$, $|s| < \delta$, with λ and $\tilde{u}$ satisfying

i) the map $s \longrightarrow (\lambda(s),\tilde{u}(s))$ from $(-\delta,\delta)$ into $\mathbb{R} \times C_o^{2,\alpha}(\bar{\Omega})$ is C^2,

ii) $\lambda(0) = \bar{\lambda}$, $\tilde{u}(0) = \bar{u}$, $\tilde{u}'(0) = v$, where v satisfies $Lv = \bar{\lambda} f_z(x,\bar{u}(x))v$, $v = 0$ on $\partial\Omega$, $v > 0$ in Ω,

iii) $\lambda'(0) = 0$, $\lambda''(0) < 0$.

It is clear that this theorem implies the existence of exactly two solutions in a deleted left neighborhood of the critical value $\bar{\lambda}$. The proof can be divided in two parts. Suppose that the curve of minimal positive solutions given by Theorem 3.4 has a limit point of the form $(\bar{\lambda},\bar{u})$ (e.g., in the topology of $\mathbb{R} \times C_o^{2,\alpha}(\bar{\Omega})$). Then, by local inversion arguments as those in Theorem 3.1 it follows that $\lim_{\lambda \uparrow \bar{\lambda}} u(\lambda) = \bar{u}$ and that the curve $u(\lambda)$ can be continuated "smoothly" through $(\bar{\lambda},\bar{u})$ "bending back" precisely in $(\bar{\lambda},\bar{u})$ in such a way that there are exactly two positive solutions in a (deleted) left neighborhood of $\bar{\lambda}$. It is possible to prove that, under assumptions (3.20)-(3.22), this is equivalent to the existence of <u>a priori</u> estimates for the solutions $u(\lambda)$, $0 \le \lambda < \bar{\lambda}$. This is not very easy for $f(x,u) = e^u$ and even more difficult for $f(x,u) = (1+\gamma u)^\beta$ (cf.[44]).

3.3. EXISTENCE OF A SECOND SOLUTION. VARIATIONAL METHODS.

Our aim in this section is to improve the results of the preceding. For that, we make some supplementary assumptions which allow the use of a different kind of methods. Under these assumptions we shall prove that (3.18),(3.19) has at least two solutions for any λ in $(0,\bar{\lambda})$. Assume that f satisfies in addition

(3.23) $$f(x,z) \le c_1 + c_2|z|^s \quad \text{for } z>0,\ 1<s<\frac{N+2}{N-2} \quad \text{if } N \ge 3 \text{ or}$$

$$f(x,z) \le e^{\psi(z)} \quad \text{with} \lim_{z\to+\infty} \frac{\psi(z)}{z^2} = 0 \quad \text{if } N=2\ ,$$

(3.24) $$\lim_{|z|\to+\infty} \frac{f(x,z)}{z} = +\infty\ ,$$

(3.25) There exist $0 \le \theta < \frac{1}{2}$ and $\bar{z} > 0$ such that for $z > \bar{z}$

$$F(x,z) = \int_0^z f(x,t)\,dt \le \theta\, z\, f(x,z)\ .$$

<u>Theorem 3.6 ([44])</u>: If f satisfies (3.20)-(3.25), then (3.18),(3.19) has at least two positive solutions for every λ in $(0,\bar{\lambda})$.

Theorem 3.6 follows from a very well-known abstract result, the "Mountain Pass Lemma", due to Ambrosetti-Rabinowitz [15] (cf. also [96], [84]).

<u>Theorem 3.7 ("Mountain Pass Lemma")</u>: Let E be a real Banach space and let $I \in C^1(E,\mathbb{R})$, with $I(0) = 0$. Suppose that I satisfies

(I_1) There exist $\alpha, \rho > 0$ such that $I(u) > 0$ in $B_\rho(0) - \{0\}$ and $I(u) \geq \alpha$ in $\partial B_\rho(0)$,

(I_2) There exists $e \in E - \{0\}$ such that $I(e) = 0$,

(I_3) (Palais-Smale condition): If (u_n) is a sequence in E such that $I(u_n) > 0$, $I(u_n)$ is bounded and $I'(u_n) \longrightarrow 0$, then u_n possesses a convergent subsequence.

Then I has a critical point $\bar{u} \in E$ such that $I(\bar{u}) \geq \alpha$.

Proof of Theorem 3.6 (sketch): We redefine f for $z < 0$ in such a way that $f > 0$ and $f(x,-z)$ satisfies (3.23)-(3.25). Then the functional

$$J(u) = \frac{1}{2} \int_\Omega |\nabla u|^2 dx - \lambda \int_\Omega F(x,u)dx$$

is $C^1(H_o^1(\Omega), \mathbb{R})$ by the continuity of f and (3.23).

We define another functional

$$I(u) = J(u + u(\lambda)) - J(u(\lambda))$$

for $0 < \lambda < \bar{\lambda}$ fixed, where $u(\lambda)$ is the minimal positive solution. Then $I(0) = 0$ and 0 is the critical point corresponding to $u(\lambda)$. The existence of a second solution follows from the application of Theorem 3.7 to the functional I, it is clear that $(I_1)-(I_3)$ will follow from assumptions (3.20)-(3.25).

COMMENTS AND BIBLIOGRAPHICAL REMARKS

For a general overwiew of nonlinear diffusion equations we send to the recent book by Smoller [106], and also to Fife [54][55] and Henry [60]. A classical reference for bifurcation theory is the book by Krasnoselski [73]. The recent book by Chow-Hale [34] covers many related topics, especially local bifurcation methods. Other useful references are the books by Berger [22], Deimling [51] and Prodi-Ambrosetti [88], the lecture notes by Iooss [66], Nirenberg [83], Rabinowitz [97] and Sattinger [102], and the surveys by Stakgold [107], Marsden [81] and Nirenberg [84]. For positive solutions there is the classical book by Krasnoselski [74] and the excellent surveys by Amann [10] and P.L. Lions [80], where many more references can be found.

Needless to say, the topological methods considered here are applicable to a very large variety of nonlinear problems. Among them, we want to mention another interesting topic, namely Ambrosetti-Prodi type problems (cf. the survey by de Figueiredo [48], and also [14][20][83]).

The iterative scheme in Section 1.1 can be found in the book by Courant-Hilbert [41]. However, the method of sub and supersolutions was only developed at the end of the sixties with papers by Cohen [36], Cohen-Laetsch [37], Keller-Cohen [70], Simpson-Cohen [105], etc (cf. the bibliography of [10]). A more general and systematic presentation was given by Amann [2] and Sattinger [100]. An abstract version of the method in the framework of ordered Banach spaces was given some years after by Amann [6][10].

As it was pointed out above, we do not try to offer a very general version of this method, but only a simplified one in order to show clearly how it works. The operator $-\Delta$ can be replaced by more general second order linear differential operators with sufficiently smooth coefficients satisfying the Maximum Principle (cf. [2][10]). The assumption that the nonlinear term f is C^1 is not necessary, the method still works if f is locally C^α $(0 < \alpha < 1)$ and there exists $M > 0$ such that for any x, $f(x,u) + Mu$ is increasing in u (cf. again [2][10]). The method is also applicable to nonlinear boundary conditions ([5][2]) (even including unilateral constraints [61]), to nonlinearities depending on the gradient (cf. [11] and its references), and to problems with discontinuous linear terms (cf., e.g., the work by Chang [33] and Stuart [109]). There are many other applications: integral and integro-differential equations, variational inequalities, etc. For all which concerns our main tool, the Maximum Principle, cf. the books by Protter-Weinberger [89] and Gilbarg-Trudinger [58].

Some interesting applications of this method to nonlinear elliptic problems can be found in the paper by Kazdan-Warner [68] (cf. also [48]). Sub and supersolutions are also applied in [29] to obtain S-shaped bifurcation curves and in [30] for the study of perturbed bifurcation problems.

The problem considered in Section 1.2 seems to arise in the theory of nuclear reactors (cf. [36][108]) and has been studied in [100][102] [61][20], etc. (Of course, many of the results in, e.g., [10] or [80] are also applicable to this problem). A basic tool are the classical results contained in the book by Courant-Hilbert [41] (cf. also [106]) concerning the properties of the eigenvalues and the eigenfunctions of (1.16)(1.17), continuous and monotone dependence on the domain and the coefficient ρ, etc. The results in this Section are taken from [61] [20] (cf. also [102][106]). The first uniqueness proof is a slight improvement of [37] and [108] (cf. the generalization in [61]), the second is included in [20]. Moreover, a fourth uniqueness proof, using Serrin's Sweeping Principle, can be found in [102]. Cf. [27] for a related result. The case of asymptotically linear f is treated in [61]. The exact multiplicity result of Proposition 1.3 was first proved in [13] by reasoning as in Section 2.2 below.

The results in Section 1.3 are contained in [62]. A first existence theorem for systems by using sub and supersolutions was already given by Sattinger [100]. Cf. [78][79][86][104], and the references therein for related results. Concerning the Maximum Principle for systems cf. [89][50]. An extension of the existence theorem in Section 1.3 was given in [53]; this generalization was motivated by the study of some free boundary problems for reaction-diffusion systems, cf. [16][52][57][64]. Cf. also [45][87] for related existence results. Uniqueness for systems is a rather difficult question: some more or less partial results can be found in, e.g., [9][45][62][78][79] and the remarks at the end of these Comments. Alternative existence proofs can be given by finding stationary points of the associated semigroups ([8][9]) or by global bifurcation arguments (cf. Section 2.7 and [82]).

We do not include here the application of these methods, mainly invariant regions, to the associated parabolic problems. Cf. the book by Smoller [106] and the work by Sattinger [100][102], Amann [8][9], Chueh-Conley-Smoller [35], Bebernes et al. [17]-[19]. For an application to a system arising in combustion theory cf. [103][60].

Concerning the stability of solutions of the stationary problem, it was proved by Sattinger [100][102] (cf. also [10]) in the case of a single equation that if the solution obtained by sub and supersolutions is unique, then it is stable and all solutions with initial data in the interval $[u_o, u^o]$ converge to this unique solution (when time goes to infinity). If there are multiple solutions, then the problem becomes more involved (cf., e.g., [101]). For stability results for systems with sub and supersolutions cf. [62][78][79][86].

A very nice presentation of topological degree theory is given in the lecture notes by Rabinowitz [97], where many applications are also included. A detailed and systematic treatment of the degree can also be found in the books by Berger [22] and Deimling [51] (cf. also [73][83][34][106]).

The results of Section 2.2 are due to Lazer-McKenna [76], cf. also [13]. For other multiplicity results cf. [10][80][14][20][48] and the corresponding bibliographies.

The global bifurcation theorems in Section 2.3 are due to Rabinowitz (cf. [91]-[93][97]), where applications to nonlinear Sturm-Liouville problems and quasilinear elliptic equations are also included. (Cf. also [83]). For bifurcation at infinity, cf. [95]. The theorems in Section 2.4 are equally due to Rabinowitz ([91][92]). Global bifurcation theorems for positive solutions were obtained independently by Dancer [46] and Turner [110], cf. also [10][12].

More information on the fixed point index is given in [10] and [85]. In particular, [10] contains a number of applications, cf. also [3] (where some results by Krasnoselski [74] are extended) and [6].

Section 2.7 is a simplified version of [63], where the case of nonlinear boundary conditions was treated. A variant of a predator-prey system considered by Conway-Smoller [39] is also studied in [63]. A related system (the Brusselator) was studied in [82] by using Theorem 2.5. As it was already pointed out in Remark 2.5 it is advantageous to employ the fixed point index, mainly because in this case it is sufficient to obtain <u>a priori</u> estimates <u>only</u> for <u>positive</u> solutions.

For other applications of the degree, cf. [20][22][48][80][83][84], etc. The Leray-Schauder degree can also be used to get stability results, cf. [99][101]. It is possible to combine the method of sub and supersolutions with the degree (or the fixed point index) to obtain multiplicity results, cf., e.g., [6][10][48][49][80].

The technique in Section 3.1 was used by Rabinowitz, cf. [90] (and [42]). The main tool (Theorem 3.1) is contained in [43], but there are similar results in the literature ([42][14][4][10]). For inversion theorems for mappings with singularities and applications cf. [13][14][88]. The theorems in Section 3.1 are in [61], cf. [4][7][10][20][69] for related work. An interesting application to a predator-prey system is given in [40]. The assumption (3.14) for the "concave" case is of the same kind of [14] or [7]. We remark that in the open problem of Fig. 6 it is possible to prove uniqueness for positive solutions in the one-dimensional case by using phase plane methods.

Sections 3.2 and 3.3 contain results by Crandall-Rabinowitz [44] (cf. also [69] for Section 3.2). These problems were treated by Guelfand [59] and Joseph-Lundgren [67] in the case of a sphere. Other results for this "forced" case are in [4][7][10]. Similar problems have been studied by Brauner-Nikolaenko (cf., e.g., [26]). For some results on S-shaped curves by using sub and supersolutions, cf. [29].

A good reference for the critical point theory used in Section 3.3 is the survey [96] by Rabinowitz, which includes some applications. The Mountain Pass Lemma is due to Ambrosetti-Rabinowitz [15], cf. also [84]. For other applications combining critical point theory with the previous methods (sub and supersolutions, degree, etc), cf. [80][94]. The critical exponent of assumption (3.23) arises also in other contexts, cf. [48][49].

Our main reason to consider nonlinear eigenvalue problems of the form

$$- \Delta u + f(u) = \lambda u \quad \text{in } \Omega ,$$
$$u = 0 \quad \text{on } \partial\Omega$$

instead of nonlinear problems of the type

$$- \Delta u = \lambda f(x,u) \quad \text{in } \Omega ,$$
$$u = 0 \quad \text{on } \partial\Omega,$$

which are more frequently studied in the literature, relies on the fact that the problem of existence of positive solutions for some reaction-diffusion systems with homogeneous Dirichlet boundary conditions can be reduced to a nonlocal perturbation of the former problem. (Needless to say, the above problems are not independent, and many of the results of, e.g., [10] apply to both of them).

More precisely, a system arising in the activator-inhibitor interaction in morphogenesis can be reduced by a decoupling technique (cf. [98][76][28]) to the equation

$$- \Delta u + Ku + f(u) = \lambda u \quad \text{in } \Omega$$
$$u = 0 \quad \text{on } \partial\Omega,$$

where K is the solution operator of a linear elliptic equation and f satisfies the usual assumptions. This problem was treated by Rothe [98] for odd f by using critical point theory. After, Lazer-McKenna [76] used degree theory for the same problem with f not necessarily odd: actually the results in Section 2.2 are the specialization of those in [76] to $K \equiv 0$.

The principal difficulty to apply the method of sub and supersolutions to this problem is the existence of some kind of Maximum Principle for the nonlocal linear operator $- \Delta + K$. A theorem of this type, together with applications to some nonlinear equations (including the previous one) was given by de Figueiredo-Mitidieri [50]. The same paper contains also an application of the Mountain Pass Lemma to related problems (cf. [71][72] for other results in the same direction). The local inversion and continuation methods in Section 3.1 can also be applied in this situation (cf. [65]), improving some of the previous results.

Another interesting example is the predator-prey system

$$\begin{aligned} -\Delta u &= au - f(u) - uv && \text{in } \Omega , \\ -\Delta v &= bu - g(u) + uv && \text{in } \Omega , \\ u &= v = 0 && \text{on } \partial\Omega , \end{aligned}$$

where $a,b > 0$ are real parameters and f and g are as above. A slight variant of this system was studied in [40], and an open problem was existence and uniqueness of nontrivial positive solutions for some range of the parameters. Existence was proved by Blat-Brown [23] by using global bifurcation and by Dancer [47], this time by using degree for cones. Here again continuation methods can be applied [65]. Related problems arising in the study of the spread of a bacterial infection (cf. [31][32]) were treated by Blat-Brown (cf. [24][25]), again by global bifurcation methods and sub and supersolutions.

REFERENCES

[1] S. Agmon, A. Douglis and L. Nirenberg. Estimates near the boundary for solutions of elliptic partial differential equations satisfying general boundary conditions, I. Comm. Pure Appl. Math. 12 (1959), 623-727.

[2] H. Amann. On the existence of positive solutions of nonlinear elliptic boundary value problems. Indiana Univ. Math. J. 21 (1971), 125-146.

[3] H. Amann. Fixed points of asymptotically linear maps in ordered Banach spaces. J. Funct. Anal. 14 (1973), 162-171.

[4] H. Amann. Multiple positive fixed points of asymptotically linear maps. J. Funct. Anal. 17 (1974), 174-213.

[5] H. Amann. Nonlinear elliptic equations with nonlinear boundary conditions. Proc. of the second Scheveningen Conf. on Differential Equations. W. Eckhaus (ed.); Amsterdam, North Holland, 1976.

[6] H. Amann. Nonlinear operators in ordered Banach spaces and some applications to nonlinear boundary value problems. *In* Nonlinear operators and the Calculus of Variations, New York, Springer, Lecture Notes in Mathematics 543, 1976, 1-55.

[7] H. Amann. Nonlinear eigenvalue problems having precisely two solutions. Math. Z. 150 (1976), 27-37.

[8] H. Amann. Invariant sets and existence theorems for semilinear parabolic and elliptic systems. J. Math. Anal. Appl. 65 (1979), 432-467.

[9] H. Amann. Existence and stability of solutions for semilinear parabolic systems, and applications to some reaction-diffusion equations. Proc. Roy. Soc. of Edinburgh 81A (1978), 35-47.

[10] H. Amann. Fixed point equations and nonlinear eigenvalue problems in ordered Banach spaces. SIAM Rev. 18 (1976), 620-709.

[11] H. Amann and M. Crandall. On some existence theorems for semilinear elliptic equations. Indiana Univ. Math. J. 27 (1978), 779-790.

[12] H. Amann and T. Laestch. Positive solutions of convex nonlinear eigenvalue problems. Indiana Univ. Math. J. 25 (1976), 259-270.

[13] A. Ambrosetti and G. Mancini. Sharp nonuniqueness results for some nonlinear problems. Nonlinear Anal. 3 (1979), 635-645.

[14] A. Ambrosetti and G. Prodi. On the inversion of some differentiable mappings with singularities between Banach spaces. Annali Mat. Pura e Appl. 93 (1973), 231-247.

[15] A. Ambrosetti and P.H. Rabinowitz. Dual variational methods in critical point theory and applications. J. Funct. Anal. 14 (1973), 349-381.

[16] C. Bandle, R.P. Sperb and I. Stakgold. Diffusion and reaction with monotone kinetics. Nonlinear Anal. 8 (1984), 321-333.

[17] J. Bebernes. A mathematical analysis of some problems from combustion theory. Quaderni dei gruppi di ricerca matematica del C.N.R., 1980.

[18] J. Bebernes, K.N. Chueh and W. Fulks. Some applications of invariance for parabolic systems. Indiana Univ. Math. J. 28 (1979), 269-277.

[19] J. Bebernes and K. Schmidt. Invariant sets and the Hukuhara-Kneser property for systems of parabolic partial differential equations. Rocky Mountain J. Math. 7 (1977), 557-567.

[20] H. Berestycki. Le nombre de solutions de certains problèmes semilinéaires elliptiques. J. Funct. Anal. 40 (1981), 1-29.

[21] H. Berestycki, B. Nikolaenko and B. Scheurer. Travelling wave solutions to reaction-diffusion systems modelling combustion. *In* Nonlinear Partial Differential Equations, Contemporary Mathematics 17, Amer. Math. Soc., 1983, 189-208.

[22] M.S. Berger. Nonlinearity and Functional Analysis. New York, Academic Press, 1977.

[23] J. Blat and K.J. Brown. Bifurcation of steady-state solutions in predator-prey and competition systems. Proc. Roy. Soc. of Edinburgh 97A (1984), 21-34.

[24] J. Blat and K.J. Brown. A reaction-diffusion system modelling the spread of bacterial infections. To appear in Math. Meth. in Appl. Sc.

[25] J. Blat and K.J. Brown. To appear.

[26] C.M. Brauner and B. Nikolaenko. Sur des problèmes aux valeurs propres non linéaires qui se prolongent en problèmes à frontière libre. C.R.A.S. Paris 288 (1979), 125-127.

[27] H. Brezis and L. Oswald. Remarks on sublinear elliptic equations. To appear.

[28] K.J. Brown. Spatially inhomogeneous steady-state solutions for systems of equations describing interacting populations. J. Math. Anal. Appl. 95 (1983), 251-264.

[29] K.J. Brown, M.M.A. Ibrahim and R. Shivaji. S-shaped bifurcation curves. Nonlinear Anal. 5 (1981), 475-486.

[30] K.J. Brown and R. Shivaji. Simple proofs of some results in perturbated bifurcation theory. Proc. Roy. Soc. of Edinburgh 93A (1982), 71-82.

[31] V. Capasso and L. Maddalena. Convergence to equilibrium states for a reaction-diffusion system modelling the spatial spread of a class of bacterial and viral diseases. J. Math. Biol. 13 (1981), 173-184.

[32] V. Capasso and L. Maddalena. Saddle point behaviour for a reaction-diffusion system: application to a class of epidemic models. Math. and Comp. Simul. XXIV (1982), 540-547.

[33] Kung-Ching Chang. On the multiple solutions of the elliptic differential equations with discontinuous nonlinear terms. Scientia Sinica 21 (1978), 139-158.

[34] S.N. Chow and J.K. Hale. Methods of Bifurcation Theory. New York, Springer, 1982.

[35] K.N. Chueh, C. Conley and J. Smoller. Positively invariant regions for systems of nonlinear diffusion equations. Indiana Univ. Math. J. 26 (1977), 373-392.

[36] D.S. Cohen. Positive solutions of nonlinear eigenvalue problems: applications to nonlinear reactor dynamics. Arch. Rat. Mech. Anal. 26 (1967), 305-315.

[37] D.S. Cohen and T. Laetsch. Nonlinear boundary value problems suggested by chemical reactor theory. J. Diff. Eq. 7 (1970), 217-226.

[38] C. Conley. Isolated invariant sets and the Morse index. Conf. Board Math. Sc. 38. Amer. Math. Soc., Providence, 1978.

[39] E. Conway and J. Smoller. Diffusion and the predator-prey interaction. SIAM J. Appl. Math. 33 (1977), 673-686.

[40] E. Conway, R. Gardner and J. Smoller. Stability and bifurcation of steady-state solutions for predator-prey equations. Adv. in Appl. Math. 3 (1982), 288-334.

[41] R. Courant and D. Hilbert. Methods of mathematical physics. New York, Interscience, 1953.

[42] M. Crandall and P. Rabinowitz. Bifurcation from simple eigenvalues. J. Funct. Anal. 8 (1971), 321-340.

[43] M. Crandall and P. Rabinowitz. Bifurcation, perturbation of simple eigenvalues and linearized stability. Arch. Rat. Mech. Anal. 52 (1973), 161-180.

[44] M. Crandall and P. Rabinowitz. Some continuation and variational methods for positive solutions of nonlinear elliptic eigenvalue problems. Arch. Rat. Mech. Anal. 58 (1975), 207-218.

[45] R. dal Passo and P. de Mottoni. Some existence, uniqueness and stability results for a class of semilinear degenerate elliptic systems. Bull. U.M.I., Anal. Funz. e Appl. 3 (1984), 203-231.

[46] E.N. Dancer. Global solutions branches for positive maps. Arch. Rat. Mech. Anal. 52 (1973), 181-192.

[47] E.N. Dancer. On positive solutions of some pairs of differential equations. Trans. Amer. Math. Soc. 284 (1984), 729-743.

[48] D.G. de Figueiredo. Lectures on boundary value problems of the Ambrosetti-Prodi type. Atas do 12º Seminario Brasileiro de Analise, Sao Paulo 1980, 230-292.

[49] D.G. de Figueiredo, P.L. Lions and R.D. Nussbaum. A priori estimates and existence of positive solutions of semilinear elliptic equations. J. Math. Pures Appl. 61 (1982), 41-63.

[50] D.G. de Figueiredo and E. Mitidieri. A maximum principle for an elliptic system and applications to semilinear problems. M.R.C. Technical Report 2653, Madison, 1984.

[51] K. Deimling. Nonlinear Functional Analysis. New York, Springer, 1985.

[52] J.I. Diaz. Nonlinear partial differential equations and free boundaries. London, Pitman, to appear.

[53] J.I. Diaz and J. Hernandez. On the existence of a free boundary for a class of reaction-diffusion systems. SIAM J. Math. Anal. 15 (1984), 670-685.

[54] P.C. Fife. Mathematical aspects of reacting and diffusing systems. New York, Springer, Lecture Notes in Biomathematics 28, 1979.

[55] P.C. Fife. Asymptotic states for equations of reaction and diffusion. Bull. Amer. Math. Soc. 84 (1978), 693-726.

[56] P.C. Fife and J.B. McLeod. The approach of solutions of nonlinear diffusion equations to travelling front solutions. Arch. Rat. Mech. Anal. 65 (1977), 335-361.

[57] A. Friedman and D. Phillips. The free boundary of a semilinear elliptic equation. Trans. Amer. Math. Soc. 282 (1984), 153-182.

[58] D. Gilbarg and N. Trudinger. Elliptic partial differential equations of second order. New York, Springer, 1977.

[59] I. Guelfand. Some problems in the theory of quasilinear equations. Amer. Math. Soc. Translations 1(2) 29 (1963), 295-381.

[60] D. Henry. Geometric theory of semilinear parabolic equations. New York, Springer, Lecture Notes in Mathematics 840, 1981.

[61] J. Hernandez. Bifurcacion y soluciones positivas para ciertos problemas de tipo unilateral. Thesis. Madrid, Universidad Autonoma, 1977.

[62] J. Hernandez. Some existence and stability results for solutions of reaction-diffusion systems with nonlinear boundary conditions. *In* Nonlinear Differential Equations: Invariance, Stability and Bifurcation, P. de Mottoni and L. Salvadori (eds.), New York, Academic Press, 1981, 161-173.

[63] J. Hernandez. Positive solutions of reaction-diffusion systems with nonlinear boundary conditions and the fixed point index. *In* Nonlinear Phenomena in Mathematical Sciences, V. Laksmikantham (ed.), New York, Academic Press, 1982, 525-535.

[64] J. Hernandez. Some free boundary problems for predator-prey systems with nonlinear diffusion. To appear in Proc. Summer School, Berkeley, 1983.

[65] J. Hernandez. In preparation.

[66] G. Iooss. Bifurcation et stabilité. Lecture notes. Université Paris-XI, Orsay, 1973.

[67] D.D. Joseph and T.S. Lundgren. Quasilinear Dirichlet problems driven by positive sources. Arch. Rat. Mech. Anal. 49 (1973), 241-269.

[68] J.L. Kazdan and F.W. Warner. Remarks on some quasilinear elliptic equations. Comm. Pure Appl. Math. 28 (1975), 567-597.

[69] J.P. Keener and H.B. Keller. Positive solutions of convex nonlinear eigenvalue problems. J. Diff. Eq. 16 (1974), 103-125.

[70] H.B. Keller and D.S. Cohen. Some positone problems suggested by nonlinear heat generation. J. Math. Mech. 16 (1967), 1361-1376.

[71] G.A. Klaasen. Stationary spatial patterns for a reaction-diffusion system with an excitable steady state. To appear.

[72] G.A. Klaasen and E. Mitidieri. Standing wave solutions for a system derived from the Fitzhugh-Nagumo equations for nerve conduction. To appear in SIAM J. Math. Anal.

[73] M. Krasnoselski. Topological methods in the theory of nonlinear integral equations. London, Pergamon Press, 1964.

[74] M. Krasnoselski. Positive solutions of operator equations. Groningen, Noordhoff, 1964.

[75] A. Ladyzenskaia and N.Uraltseva. Linear and quasilinear elliptic equations. New York, Academic Press, 1968.

[76] A.C. Lazer and P.J. McKenna. On steady-state solutions of a system of reaction-diffusion equations from biology. Nonlinear Anal. 6 (1982), 523-530.

[77] J. Leray and J. Schauder. Topologie et équations fonctionnelles. Ann. Sci. Ecole Norm. Sup. 51 (1934), 45-78.

[78] A. Leung. Monotone schemes for semilinear elliptic systems related to ecology. Math. Meth. in the Appl. Sci. 4 (1982), 272-285.

[79] A. Leung and D. Clark. Bifurcations and large-time asymptotic behavior for prey-predator reaction-diffusion equations with Dirichlet boundary data. J. Diff. Eq. 35 (1980), 113-127.

[80] P.L. Lions. On the existence of positive solutions of semilinear elliptic equations. SIAM Rev. 24 (1982), 441-467.

[81] J.E. Marsden. Qualitative methods in bifurcation theory. Bull. Amer. Math. Soc. 84 (1978), 1123-1148.

[82] G. Meurant and J.C. Saut. Bifurcation and stability in a chemical system. J. Math. Anal. Appl. 59 (1977), 69-92.

[83] L. Nirenberg. Topics in Nonlinear Functional Analysis. Lecture notes. New York, Courant Institute, 1974.

[84] L. Nirenberg. Variational and topological methods in nonlinear problems. Bull. Amer. Soc. 4 (1981), 267-302.

[85] R.D. Nussbaum. The fixed point index for local condensing maps. Annali Mat. Pura e Appl. 89 (1971), 217-258.

[86] C.V. Pao. On nonlinear reaction-diffusion systems. J. Math. Anal. Appl. 87 (1982), 165-198.

[87] M.A. Pozio and A. Tesei. Degenerate parabolic problems in population dynamics. To appear.

[88] G. Prodi and A. Ambrosetti. Analisi non lineare, Quaderno I. Pisa, Scuola Normale Superiore, 1973.

[89] M. Protter and H.F. Weinberger. Maximum Principles in Differential Equations. Prentice-Hall, Englewood Cliffs, 1967.

[90] P.H. Rabinowitz. A note on a nonlinear eigenvalue problem for a class of differential equations. J. Diff. Eq. 9 (1971), 536-548.

[91] P.H. Rabinowitz. Some global results for nonlinear eigenvalue problems. J. Funct. Anal. 7 (1971), 487-513).

[92] P.H. Rabinowitz. A global theorem for nonlinear eigenvalue problems and applications. *In* Contributions to Nonlinear Functional Analysis, E.H. Zarantonello (ed.), New York, Academic Press, 1971, 11-36.

[93] P.H. Rabinowitz. Some aspects of nonlinear eigenvalue problems. Rocky Mountain J. Math. 3 (1973), 161-202.

[94] P.H. Rabinowitz. Pairs of positive solutions of nonlinear elliptic partial differential equations. Indiana Univ. Math. J. 23 (1973), 173-186.

[95] P.H. Rabinowitz. On bifurcation from infinity. J. Diff. Eq. 14 (1973), 462-475.

[96] P.H. Rabinowitz. Variational methods for nonlinear eigenvalue problems. *In* Eigenvalues of nonlinear problems, G. Prodi (ed.), Roma, Edizioni Cremonese, 1974, 141-195.

[97] P.H. Rabinowitz. Théorie du degré topologique et applications dans les problèmes aux limites non linéaires. Lecture notes by H. Berestycki. Université Paris-VI, 1975.

[98] F. Rothe. Global existence of branches of stationary solutions for a system of reaction-diffusion equations from biology. Nonlinear Anal. 5 (1981), 487-498.

[99] D.H. Sattinger. Stability of bifurcating solutions by Leray-Schauder degree. Arch. Rat. Mech. Anal. 43 (1971), 154-166.

[100] D.H. Sattinger. Monotone methods in nonlinear elliptic and parabolic equations. Indiana Univ. Math. J. 21 (1972), 979-1000.

[101] D.H. Sattinger. Stability of solutions of nonlinear equations. J. Math. Anal. Appl. 39 (1972), 1-12.

[102] D.H. Sattinger. Topics in Stability and Bifurcation Theory. New York, Springer, Lecture Notes in Mathematics 309, 1973.

[103] D.H. Sattinger. A nonlinear parabolic system in the theory of combustion. Quarterly J. Appl. Math. 33 (1975), 47-62.

[104] A. Schiaffino and A. Tesei. Competition systems with Dirichlet boundary conditions. J. Math. Biol. 15 (1982), 93-105.

[105] R.B. Simpson and D.S. Cohen. Positive solutions of nonlinear elliptic eigenvalue problems. J. Math. Mech. 19 (1970), 895-910.

[106] J. Smoller. Shock waves and reaction-diffusion equations. New York, Springer, 1983.

[107] I. Stakgold. Branching of solutions of nonlinear equations. SIAM Rev. 13 (1971), 289-332.

[108] I. Stakgold and L. Payne. Nonlinear problems in nuclear reactor analysis. *In* Nonlinear Problems in Physics and Biology, New York, Springer, Lecture Notes in Mathematics 322, 1973, 298-309.

[109] C.A. Stuart. Maximal and minimal solutions of elliptic differential equations with discontinuous nonlinearities. Math. Z. 163 (1978), 239-249.

[110] R.E.L. Turner. Transversality and cone maps. Arch. Rat. Mech. Anal. 58 (1975), 151-179.

REACTION-DIFFUSION PROBLEMS IN CHEMICAL ENGINEERING

Ivar Stakgold
Department of Mathematical Sciences
University of Delaware
Newark, Delaware 19716

Introduction

In this set of lectures, we shall investigate some reaction-diffusion problems arising in chemical engineering. The range of phenomena that can occur is quite remarkable and we shall confine ourselves to a few problems that illustrate the rich variety of the field. Some important areas such as periodic oscillations and flame propagation will not be touched upon at all. Among the important books that deal with reaction-diffusion problems in chemical engineering, we mention here those of Aris [1], Buckmaster and Ludford [7], and Bischoff and Froment [14]. I have also relied on a number of valuable review papers, including those of Aris [2], Chandra and Davis [9], Boddington, Gray and Wake [6], and Pao [19]. Additional references on specific points will appear in the text.

1. The basic equations and some elementary applications

The basic equations governing chemical reactions are formulations of the conservation of mass and heat. The conservation of scalar physical quantities can be expressed in general as follows. Let Ω be a domain with boundary $\partial\Omega$ and let D be a bounded subdomain of Ω with boundary B. Let $H_D(t)$ be the accumulation or total amount in D at time t of the physical quantity under consideration; let $G_D(t)$ be the production in D per unit time, that is the amount of the quantity being generated (perhaps through a chemical process) per unit time in D at time t; let $F_D(t)$ be the amount of the quantity flowing into D through B per unit time.

Our basic conservation equation then takes the form

$$\frac{dH_D}{dt} = G_D + F_D. \tag{1.1}$$

We shall assume that G_D and H_D can be expressed as densities

$$G_D(t) = \int_D g(x,t)\, dx , \quad H_D(t) = \int_D h(x,t)\, dx .$$

We shall also assume that F_D can be written as a surface integral of the inward component of a flux density vector $\vec{f}$. Thus,

$$F_D(t) = - \int_B \vec{f} \cdot \vec{n}\, ds ,$$

where $\vec{n}$ is the outward normal to D. Using the divergence theorem, we see that (1.1) reduces to

$$\int_D \left(\frac{\partial h}{\partial t} - g + \operatorname{div} \vec{f} \right) dx = 0 ,$$

for <u>all</u> subdomains in Ω. The postulated continuity of the integrand then yields

$$\frac{\partial h}{\partial t} + \operatorname{div} \vec{f} = g , \quad \text{for all } x \text{ in } \Omega \text{ and all } t . \tag{1.2}$$

Let us now consider the mass and heat balance for a species of gas that is both diffusing and reacting. First we look at the mass balance, setting $h = c$, the concentration of the species in moles/cm^3. The production term g is a function of c and the temperature T. If c is the concentration of a reactant in an irreversible reaction, g will be negative and will depend on c and the absolute temperature T through the Arrhenius law

$$g = - Ac^m e^{-E/RT} , \tag{1.3}$$

where E is the activation energy, R the universal gas constant, m the order of the reaction and A the preexponential factor. Here E, R, A are positive, $m \geq 0$ and $g = 0$ when $c = 0$. The reaction stops when there is no reactant! The quantity E/R is in units of temperature so that E/RT is dimensionless. Let $s(T) = \exp(-E/RT)$; then s is defined on $(0,\infty)$ and is strictly increasing with $s(0+) = 0$, $s(\infty) = 1$. Furthermore, s is convex on $\left(0, \frac{E}{2R}\right)$ and concave for $T > \frac{E}{2R}$, with $\left(\frac{E}{2R}, e^{-2}\right)$ the only inflection point.

Assuming no transport, the diffusive flux may be written as

$$\vec{f} = - D \operatorname{grad} c ,$$

where D is the diffusivity in cm^2/sec. For simplicity, we shall suppose that D is constant. Equation (1.2) thus becomes

$$\frac{\partial c}{\partial t} - D\, \Delta c = - Ac^m e^{-E/RT} , \tag{1.4}$$

where Δ is the Laplacian operator.

Next let us consider the heat balance. The quantity h is now the enthalpy (in calories/cm^3) whose rate of change is given by $C \frac{\partial T}{\partial t}$, where C is the specific heat of the mixture per unit volume. We assume that C is a constant. The production of heat will be proportional to the consumption of reactant. Thus the production term takes the form

$$QAc^m e^{-E/RT},$$

where Q is the exothermicity per mole (note that $Q < 0$ for an endothermic reaction and zero for an isothermal reaction). By Fourier's law, the flux of heat is $-k \operatorname{grad} T$, where k is the thermal conductivity, regarded here as constant. Therefore (1.2), in this setting, becomes

$$C \frac{\partial T}{\partial t} - k \Delta T = QAc^m e^{-E/RT} \tag{1.5}$$

We regard (1.4) and (1.5) as a system of nonlinear equations for c and T. The initial conditions are $c(x,0) = c_0(x)$ and $T(x,0) = T_0(x)$. What are the boundary conditions associated with the problem? If the reaction takes place in a closed vessel such that no reactant can enter or leave the vessel after $t = 0$, the appropriate boundary condition is

$$\frac{\partial c}{\partial \nu} = 0 \;, \quad x \in \partial\Omega \;, \quad t > 0 \;, \tag{1.6}$$

where ν is the outer normal. The temperature T might satisfy the condition

$$T = T_f \quad \text{or} \tag{1.7a}$$

$$-k \frac{\partial T}{\partial \nu} = \chi(T-T_f) \;, \tag{1.7b}$$

where T_f is the constant ambient temperature and χ is the Newtonian heat transfer coefficient at the boundary. One can regard (1.7a) as the limit as $\chi \to \infty$ of (1.7b).

When dealing with a reaction within a catalyst particle, the same boundary condition (1.7a) or (1.7b) applies to the temperature. But now reactant enters the particle from a rich environment to sustain the reaction and therefore the appropriate condition on c is

$$c = c_f \tag{1.8}$$

or a condition on c of type (1.7b) which we shall not consider here.

Problem (1.4), (1.5), (1.6), (1.7a) or (1.7b) will be called the <u>combustion</u> problem. Problem (1.4), (1.5), (1.7a), (1.8) will be known as the <u>catalyst</u> problem, where it is of course understood that the constant D in (1.4) is now an effective diffusivity for the porous catalyst particle. Initial conditions are associated with both problems.

There are spatially homogeneous problems that are analogous to the combustion and catalyst problems. If conditions are uniform throughout the combustion vessel, the Laplacians in (1.4) and (1.5) drop out. The initial conditions are taken to be constants. Boundary condition (1.6) is consistent with spatial homogeneity. Newtonian cooling can no longer be represented by (1.7a) or (1.7b) but can be incorporated as a loss term in the differential equation for T. This yields the system

$$\frac{dc}{dt} = -Ac^m e^{-E/RT}, \quad c(0) = c_0 \tag{1.9}$$

$$C\frac{dT}{dt} = -\frac{\chi S}{V}(T-T_f) + QAc^m e^{-E/RT}, \quad T(0) = T_0, \tag{1.10}$$

where S/V is the surfact to volume ratio of the vessel. For simplicity in the analysis, we will usually take $T_0 = T_f$. Problem (1.9), (1.10) will be called the lumped combustion problem.

If $\chi > 0$, the only possible steady state for (1.9), (1.10) is $c = 0$, $T = T_f$. If $\chi = 0$, multiply (1.9) by Q and add to (1.10) to obtain

$$Qc + CT = Qc_0 + CT_0 .$$

As $t \to \infty$, $c \to 0$ so that T tends to $T_0 + Qc_0/C$, where Qc_0/C is the adiabatic temperature rise.

Another quantity of importance in the sequel is the characteristic reaction time t_R which is the time required to complete the reaction if it is continued at the initial rate. Solving (1.9) with $T = T_0$, $c = c_0$, we find

$$t_R = e^{E/RT_0}/Ac_0^{m-1} \tag{1.11}$$

As pointed by Aris [1], the stirred reactor whose equations we derive below bears similarities with the catalyst particle. Spatial conditions are uniform in the stirred reactor. A stream of reactant concentration c_f and temperature T_f feeds the reactor at a flow rate q, and the products are removed after reaction at the same flow rate. The reactant concentration c and temperature T in the reactor are the same as those of the product stream

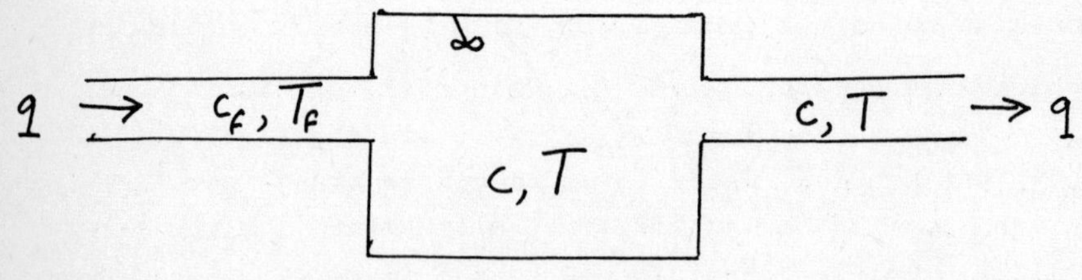

Fig. 1

The equations for the reactor are

$$\frac{dc}{dt} = \frac{q}{V}(c_f - c) - Ac^m e^{-E/RT}, \quad c(0) = c_f \tag{1.12}$$

$$C\frac{dT}{dt} = \frac{qC}{V}(T_f - T) + QAc^m e^{-E/RT}, \quad T(0) = T_f . \qquad (1.13)$$

In (1.13), we have neglected the heat capacity of the walls and have chosen favorable initial conditions. Clearly (1.13) and (1.10) are the same equation; (1.12) permits reactant to enter the system whereas (1.9) will exhaust the reactant either in finite time or as $t \to \infty$. Thus (1.12) and (1.13) are analogous to the catalyst problem (1.4), (1.5), (1.7a), (1.8). We call (1.12), (1.13) the stirred reactor problem.

For small values of q, (1.11) (1.12) has a unique steady state c_s, T_s which can be calculated. For larger values of q, there can be three or one steady states.

Nondimensionalization. We begin with the spatially homogeneous cases (1.9), (1.10) and (1.12), (1.13). In (1.10) we take $T_0 = T_f$ and, in order to handle both the stirred reactor and lumped combustion problems simultaneously, we relabel c_0 in (1.9) as c_f.

We introduce a nondimensional concentration u and temperature v as follows:

$$u = c/c_f , \quad v = (T - T_f)/\varepsilon T_f \qquad (1.14)$$

where $\varepsilon = RT_f/E$ is a dimensionless reciprocal activation energy. In many problems, ε is small. In any event $e^{-E/RT}$ can be written as $e^{-1/\varepsilon}\ e^{v/(1+\varepsilon v)}$. The stirred reactor problem then takes the form

$$\frac{du}{dt} = -\frac{(u-1)}{\tau} - \frac{u^m}{t_R} e^{v/(1+\varepsilon v)}, \quad u(0) = 1$$

$$\frac{dv}{dt} = -\frac{v}{\tau} + \frac{B}{t_R} u^m e^{v/(1+\varepsilon v)}, \quad v(0) = 0$$

where $\tau = V/q$ is the residence time, t_R is given by (1.11) and $B = Qc_f/C\varepsilon T_f$ is a dimensionless adiabatic temperature rise (actual adiabatic rise divided by εT_f). Note that $v = 0$ now represents the feed and initial temperature.

The lumped combustion problem becomes

$$\frac{du}{dt} = -\frac{u^m}{t_R} e^{v/(1+\varepsilon v)}, \quad u(0) = 1$$

$$\frac{dv}{dt} = -a\,v + \frac{B}{t_R} u^m e^{v/(1+\varepsilon v)}, \quad v(0) = 0 ,$$

where $a = \chi S/VC$ is a nondimensional reciprocal thermal relaxation time.

It remains to nondimensionalize the time. For the stirred reactor it is natural to use t/τ as the nondimensional time, whereas for the combustion problem t/t_R is the accepted choice. We shall use t/t_R

in both problems and then label the new time by the same letter t as the dimensional time. We are thus led to the final nondimensional systems.

$$\text{(S.R.)}\quad \begin{cases} \dfrac{du}{dt} = -\, b(u-1) - u^m e^{v/(1+\varepsilon v)} & u(0) = 1 \\[2mm] \dfrac{dv}{dt} = -\, b\, v + B\, u^m e^{v/(1+\varepsilon v)} & v(0) = 0 \end{cases}$$

$$\text{(L.C.)}\quad \begin{cases} \dfrac{du}{dt} = -\, u^m e^{v/(1+\varepsilon v)} & u(0) = 1 \\[2mm] \dfrac{dv}{dt} = -\, b\, v + B\, u^m e^{v/(1+\varepsilon v)} & v(0) = 0\ . \end{cases}$$

There are three parameters ε, B, b in the two problems: ε, B have the same significance in both problems but $b = a\, t_R$ in L.C., whereas $b = t_R/\tau$ in S.R.

Nondimensionalization proceeds in a similar way for the reaction-diffusion equations (1.4), (1.5). We scale the time variable, the concentration and the temperature as before, but we must also introduce a space scaling $x' = x/d$, where d is a typical dimension of Ω. We then obtain the pair of equations (after dropping the prime on x)

$$\frac{\partial u}{\partial t} - \mu\, \Delta u = -\, u^m e^{v/(1+\varepsilon v)} \tag{1.15}$$

$$\sigma \frac{\partial v}{\partial t} - \Delta v = \delta\, u^m e^{v/(1+\varepsilon v)} \tag{1.16}$$

where $\mu = D\, t_R/d^2$, $\sigma = C\, d^2/k\, t_R$ and $\delta = B\sigma$.

We shall consider two boundary value problems associated with (1.1). In the <u>distributed combustion problem</u> (D.C.) the boundary conditions will be taken as

$$\frac{\partial u}{\partial \nu} = 0\ , \quad v = 0 \quad \text{for} \quad x \in \partial\Omega\ , \quad t > 0, \tag{1.17}$$

whereas in the <u>catalyst</u> problem (CAT.)

$$u = 1\ , \quad v = 0 \quad \text{for} \quad x \in \partial\Omega\ , \quad t > 0. \tag{1.18}$$

The initial conditions for both types of problems will usually be

$$u(x,0) = 1\ , \quad v(x,0) = 0 \tag{1.19}$$

Let us now examine the lumped combustion problem

$$\frac{du}{dt} = -\, u^m e^{v/(1+\varepsilon v)}\ , \qquad u(0) = 1 \tag{1.20}$$

$$\frac{dv}{dt} = -\, bv + Bu^m e^{v/(1+\varepsilon v)}\ , \quad v(0) = 0\ . \tag{1.21}$$

Taking $B \geq 0$ (isothermal or exothermic reaction), we see that

$v \geq 0$. The function $f(v) = e^{v/(1+\varepsilon v)}$ is increasing on $[0,\infty)$ with $f(0) = 1$, $f(\infty) = e^{1/\varepsilon}$. If $\varepsilon \geq 1/2$, the function is concave for $v > 0$. If $\varepsilon < 1/2$, the function is convex for $v < (1-2\varepsilon)/\varepsilon^2$ and then concave.

If $Q = 0$, then $B = 0$ and $v \equiv 0$ so that the process is isothermal and

$$\frac{du}{dt} = - u^m , \quad u(0) = 1 .$$

Obviously $u \to 0$ as $t \to \infty$. If $m < 1$, then $u \equiv 0$ for $t \geq 1/(1-m)$.

In the exothermic case $(Q>0)$, we have $v(t) > 0$ for $t > 0$ and therefore the reactant is consumed at a greater rate than in the isothermal case and hence the concentration $u(t)$ is smaller. Thus, we still have $u(t) \to 0$ as $t \to \infty$. If $b > 0$, we can easily show that $v(t) \to 0$ as $t \to \infty$. Indeed, since $u^m \to 0$, (1.21) becomes

$$\frac{dv}{dt} + bv \leq \alpha(t) , \quad v(0) = 0 ,$$

where $\alpha(t) \to 0$ as $t \to \infty$. Thus we obtain

$$v(t) \leq e^{-bt} \int_0^t \alpha(s)\, e^{bs}\, ds .$$

Given a small positive θ we pick T so large that $\alpha < \theta$ for $t > T$. Splitting the interval of integration into two parts from 0 to T and T to t, we find

$$v(t) \leq A(T)\, e^{-bt} + \frac{\theta}{b} \quad \text{for all} \quad t > T.$$

Now choose t sufficiently large so that $e^{-bt} \leq \frac{\theta}{A(T)}$. The desired result then follows. The simple steady state $u = 0$, $v = 0$ is asymptotically stable. What then is the question of interest? If ε is small there can be a large temperature rise before the temperature settles down to moderate values. This rise in temperature can be identified with ignition. We note that by multiplying (1.20) by B and adding to (1.21)

$$\frac{d}{dt}(Bu+v) \leq 0$$

and hence $v \leq B$. Therefore the temperature excursion cannot exceed B. For a careful asymptotic analysis of the ignition problem for small ε, see [17] or [18].

There are two approximations popular in combustion theory. If the heat of reaction is large, the temperature variations take place on a much shorter time scale than the reactant depletion. It is therefore reasonable to neglect reactant consumption and to rescale the time in (1.21) to obtain

$$\frac{dv}{dt} = -\beta v + e^{v/(1+\varepsilon v)}, \quad v(0) = 0. \tag{1.22}$$

where $\beta = b/B$.

A second type of approximation in based on high activation energy ($\varepsilon << 1$). In that case one can replace the nonlinear temperature term in (1.22) or in (1.20), (1.21) by e^v. For (1.22) this would yield

$$\frac{dv}{dt} = -\beta v + e^{v}, \quad v(0) = 0 . \tag{1.23}$$

We have already seen that the temperature in (1.20), (1.21) cannot exceed B and that information should be used as a constraint in the approximations (1.22) and (1.23). This would, however, make the calculations awkward and we shall not take the known upper bound into account.

In (1.23) if $\beta < e$, there is no steady state. The losses are so small that the time-dependent solution blows up in finite time. If $\beta = e$ there is one steady state and if $\beta > e$ there are two steady states, the smaller of which is approached from below as $t \to \infty$.

The behavior of (1.22) is similar in many ways to the stirred reactor problem

$$\frac{du}{dt} = -b(u-1) - u^m e^{v/(1+\varepsilon v)} \qquad u(0) = 1 \tag{1.24}$$

$$\frac{dv}{dt} = -bv + Bu^m e^{v/(1+\varepsilon v)} \qquad v(0) = 0. \tag{1.25}$$

Multiply (1.24) by B and add to (1.25) to obtain

$$\frac{d}{dt}(Bu+v) = -b(Bu+v) + bB ,$$

so that, in view of the initial conditions,

$$Bu + v \equiv B$$

and

$$v = B(1-u) \quad \text{or} \quad u = 1 - (v/B).$$

Note therefore that, since $u \geq 0$, the temperature cannot exceed B, the dimensionless adiabatic temperature rise. Substituting in (1.24) we obtain a scalar equation for the temperature v:

$$\frac{dv}{dt} = -\beta v + [1 - (v/B)]^m e^{v/(1+\varepsilon v)}, \quad v(0) = 0, \tag{1.26}$$

where the time has been rescaled in (1.22). The term in brackets on the right side of (1.26) is interpreted to be zero if $v \geq B$.

We shall analyze (1.22) and (1.26) simultaneously. To be explicit and keep the calculations reasonable we will only consider the case

$m = 1$ in (1.26). We can think of (1.22) as a limit as $B \to \infty$ of (1.26).

The possible steady states can be found by setting the right side of (1.26) equal to zero. These steady states are then the intersections of the horizontal line $y = \beta$ and of the curve $y = F(v) = \left(\frac{1}{v} - \frac{1}{B}\right) e^{v/(1+\varepsilon v)}$.

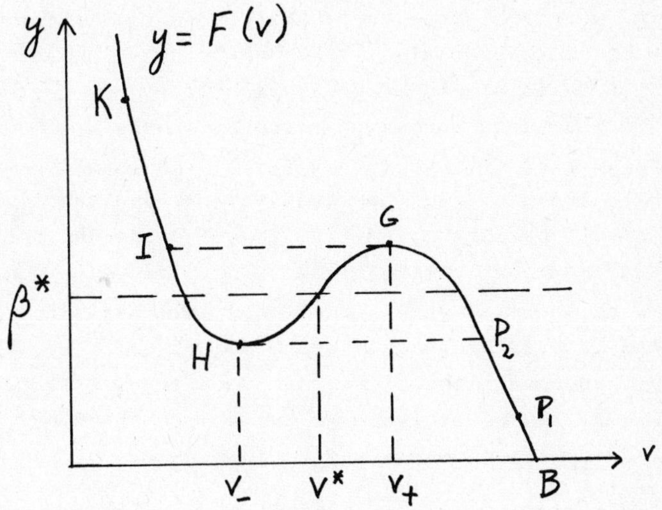

Fig. 2

Clearly $F(0+) = \infty$, $F(B) = 0$ and a straightforward calculation gives

$$F'(v) = -P(v)\left[\left(\frac{1}{B} + \varepsilon^2\right) v^2 + (2\varepsilon - 1) v + 1\right], \text{ where } P > 0.$$

The discriminant of the quadratic term is $1 - 4\varepsilon - \frac{4}{B}$. If, therefore, $\varepsilon > \frac{1}{4} - \frac{1}{B}$, F is strictly decreasing and there is exactly one steady state. Since $\varepsilon > 0$ the inequality is satisfied for all $B \leq 4$. If $\varepsilon > \frac{1}{4}$ the inequality is satisfied for all B.

The interesting case is when $B > 4$ and $\varepsilon < \frac{1}{4} - \frac{1}{B}$. Then $F(v)$ has the appearance shown in Figure 2. If $\beta > F(v_+)$ or $\beta < F(v_-)$ there is one steady-state solution. If $F(v_-) < \beta < F(v_+)$ then there are three steady-state solutions. It is easily seen that the states on the middle part of the curve (between v_- and v_+) are unstable. Indeed suppose $\beta = \beta^*$ and let the central solution be v^*. If we consider the initial problem (1.26) with $v(0)$ different slightly from v^* (and $m = 1$), we see that the right side of (1.26) is of the same sign as $v(0) = v^*$ and will therefore tend to move away from v^* as t increases.

The states on the other parts of the curve are seen to be stable.

We recall that β is proportional to the flow rate for the stirred reactor. We can trace the evolution of the steady state as β is slowly increased. For low flow rate the steady state is represented by a point such as P_1. As we increase the flow rate we move upward along the branch between B and v_+. When we reach G an increase in flow rate forces us to jump to the stable point I (this corresponds to a quenching of the reaction since the temperature at I is low and there is little conversion to products). Increasing the flow rate further makes us more upward to such a state as K. If we now decide to decrease the flow rate we move down through the stable states on the left branch of the curve all the way to H when ignition takes place and we jump to the state P_2 at much higher temperature. We can handle the combustion problem (1.22) by letting $B \to \infty$ in the preceding analysis. Now $F(v) > 0$ for all v and $\lim_{v\to\infty} F(v) = 0$. The picture is similar to that in Figure 2 , with the tail of the curve now asymptotic to the v axis. If $\varepsilon > \frac{1}{4}$ there is a single steady state and it is stable. If $\varepsilon < \frac{1}{4}$ and β large (large heat losses) we are operating at low temperature at a point like K. As we decrease β we descend the left branch until we reach H when ignition occurs and we jump to P_2.

For problem (1.23), obtained from (1.22) by setting $\varepsilon = 0$, $F(v) = \frac{1}{v} e^v$ whose single critical point $v = 1$ yields an absolute minimum for F. As v increases from 0 to 1, F decreases from ∞ to e; as v increases from 1 to ∞, F increases from e to ∞. There are therefore two steady states if $\beta > e$ and none if $\beta < e$. If $\beta > e$ the smaller of the steady states is stable. If $\beta > e$, the solution of the initial value problem (1.23) blows up in finite time. We have, from (1.23),

$$\frac{dv}{e^v - \beta v} = dt \ , \quad \text{or} \quad t = \int_0^v \frac{dz}{e^z - \beta z} \ .$$

For $\beta < e$, $K = \int_0^\infty dz/(e^z - \beta z)$ is finite so that $v \to \infty$ as $t \to K$.

2. Distributed combustion and the catalyst particle

In this section we shall consider problem (1.15), (1.16) with initial conditions (1.19) and either the boundary conditions (1.17) for the combustion problem or (1.18) for the catalyst particle.

From the maximum principle for parabolic equations, it is clear that the concentration $u(x,t)$ satisfies $u < 1$ in $\Omega \times (0,\infty)$. Therefore the temperature v satisfies the inequality

$$v_t - \frac{1}{\sigma}\Delta v \le \frac{\delta}{\sigma} e^{1/\varepsilon} \; ; \quad v(x,0) = 0 , \quad v(\partial\Omega,t) = 0 .$$

Let us set

$$v = \delta e^{1/\varepsilon} w + z ,$$

where $w(x)$ satisfies the Poisson equation

$$- \Delta w = 1 , \quad x \in \Omega , \quad w(\partial\Omega) = 0 . \tag{2.1}$$

We then find that

$$z_t - \frac{1}{\sigma}\Delta z \le 0 \; ; \quad z(x,0) \le 0 \qquad z(\partial\Omega,t) = 0 ,$$

so that $z \le 0$ for all x, t. Therefore $v(x,t)$ is uniformly bounded in space-time.

If the diffusivities $\mu, \frac{1}{\sigma}$ in (1.15) and (1.16) are equal, we can combine the two equations to obtain, for $h = \delta\mu u + v$,

$$h_t - \mu\Delta h = 0 , \quad h(x,0) = \delta\mu , \quad h(\partial\Omega,t) \le \delta\mu . \tag{2.2}$$

In the catalyst particle the last inequality is replaced by an equality so that $h \equiv \delta\mu$ for all x and t, and, hence

$$u = 1 - \frac{v}{\delta\mu} , \quad v = \delta\mu(1-u) . \tag{2.3}$$

Since $u \ge 0$, $v \le \delta\mu$ for all x, t. The system (1.15), (1.16) is then reduced to the scalar problem

$$\sigma v_t - \Delta v = \delta\left(1 - \frac{\sigma v}{\delta}\right)^m e^{v/(1+\varepsilon v)} \; ; \quad x \in \Omega , \quad t > 0 \; ;$$
$$v(x,0) = 0 , \quad v(\partial\Omega,t) = 0 . \tag{2.4}$$

For the combustion problem with equal diffusivities, the inequality in (2.2) must be preserved and therefore, by the maximum principle,

$$h(x,t) \le \delta\mu$$

and

$$\delta\mu u + v \le \delta\mu , \quad v \le \delta\mu(1-u) . \tag{2.5}$$

Returning to the case of unequal diffusivities, we can show that $u(x,t)$ and $v(x,t)$ both tend to 0 as $t \to \infty$ for the combustion problem. As in the spatially homogeneous case, the interest is to analyze the large rise in temperature that may occur before the approach to the steady state. Existence and uniqueness can be proved by using the method of upper and lower solutions (see [19]) or by comparison theorems (see [10]). We now consider some of the approximations used for the combustion problem. As in the lumped problem we can neglect (for Q large) reactant consumption; this leads to the temperature equation

$$\sigma v_t - \Delta v = \delta e^{v/(1+\varepsilon v)} \;;\; v(x,0) = v(\partial\Omega,t) = 0 \;. \tag{2.5}$$

In making our approximation we have not taken into account the fact that v is known, a priori, to be bounded. This should be added as a constraint on (2.5), but it is simpler computationally to disregard this upper bound.

The steady state equation is given by

$$-\Delta v = \delta e^{v/(1+\varepsilon v)} \;,\; x \in \Omega \;;\; v(\partial\Omega) = 0 \;. \tag{2.6}$$

We shall prove existence by the method of upper and lower solutions. A function $\underline{v}(x)$ is a <u>lower solution</u> (L.S.) to (2.6) if it satisfies

$$-\Delta \underline{v} \le \sigma e^{\underline{v}/(1+\varepsilon \underline{v})} \;,\; x \in \Omega \;;\; \underline{v}(\partial\Omega) \le 0 \;. \tag{2.7}$$

Similarly $\overline{v}$ is an upper solution (U.S.) if it satisfies (2.7) with <u>both</u> inequalities reversed. We quote the following theorem (see [ ], for instance). If one can find a L.S. $\underline{v}(x)$ and an U.S. $\overline{v}(x)$ such that $\underline{v}(x) \le \overline{v}(x)$, then there exists at least one solution of (2.6) satisfying $\underline{v}(x) \le v(x) \le \overline{v}(x)$. Since our nonlinearity is bounded below by δ and above by $\delta e^{1/\varepsilon}$, we can take $\underline{v} = \delta w$, $\overline{v} = \delta e^{1/\varepsilon} w$, where w is the solution of (2.1). Thus, there exists at least one solution $v(x)$ of (2.6), and every positive solution of (2.6) satisfies

$$\delta w(x) \le v(x) \le \delta e^{1/\varepsilon} w(x) \;.$$

Uniqueness is a more delicate question. Although necessary and sufficient conditions are not available, it is easy to find a sufficient condition.

<u>Theorem</u>. Let $f(v) > 0$ for $v > 0$ and let $\frac{f(v)}{v}$ be strictly decreasing for $v > 0$. Then the boundary value problem

$$-\Delta v = f(v) \;,\; x \in \Omega \;;\; v(\partial\Omega) = 0 \;; \tag{2.8}$$

has at most one solution.

<u>Proof</u>. Let v_1 and v_2 be two positive solutions with $v_2(x) > v_1(x)$ over the subdomain D. On the boundary ∂D of D (which may, in part, coincide with $\partial\Omega$), $v_1 = v_2$ and $\frac{\partial v_2}{\partial \nu} \le \frac{\partial v_1}{\partial \nu}$ where ν is the outward normal to ∂D. Combining the equations for v_1 and v_2 in the usual way and integrating over D, we find

$$\int_{\partial D} \left(v_1 \frac{\partial v_2}{\partial \nu} - v_2 \frac{\partial v_1}{\partial \nu} \right) ds = - \int_D [v_1 f(v_2) - v_2 f(v_1)] dx$$

$$= \int_D v_1 \, v_2 \left[\frac{f(v_1)}{v_1} - \frac{f(v_2)}{v_2} \right] dx \ .$$

Now the surface integral is nonpositive, whereas the right side is positive. The contradiction proves uniqueness.

For the particular case (2.6) we have shown that $\frac{1}{v} e^{v/(1+\varepsilon v)}$ is strictly decreasing if $\varepsilon \geq \frac{1}{4}$. Therefore we have uniqueness in that range. For smaller values of ε we may have a multiplicity of solutions. An exact count of solutions is difficult except in low-dimensional spaces (n=1,2). In any event, the solution $v(x,t)$ of (2.5) tends as $t \to \infty$ to the minimal solution of the steady problem.

Let us now consider the Gelfand approximation in which ε is set equal to zero in (2.6):

$$- \Delta v = \delta e^{v} \ , \quad x \in \Omega \ ; \quad v(\partial\Omega) = 0 \ . \tag{2.9}$$

Since e^{v} is not bounded, our previous existence proof fails. In fact we can show that for δ large enough, (2.9) has no solution. To prove this, consider the linear eigenvalue problem

$$- \Delta \phi = \lambda \phi \ , \quad x \in \Omega \ ; \quad \phi(\partial\Omega) = 0 \ , \tag{2.10}$$

whose lowest eigenvalue λ_1 is known to be positive with corresponding eigenfunction of one sign in Ω. We choose the eigenfunction to be positive in Ω with $\max \phi_1 = 1$. Combining with (2.9), integrating over Ω and using the boundary conditions yields

$$\int_\Omega \phi_1 (\delta e^{v} - \lambda_1 v) dx = 0 \ .$$

Since $\phi_1 > 0$ in Ω, a solution is possible only if the curves $y = e^{v}$ and $y = \frac{\lambda_1}{\delta} v$ intersect for $v > 0$. There is a critical value of δ, say δ^*, beyond which the curves do not intersect. This value is determined by a tangency condition so that both $e^{v} = \frac{\lambda_1}{\delta} v$ and $e^{v} = \frac{\lambda_1}{\delta}$. Hence $v = 1$ and $\delta^* = \lambda_1/e$. Thus (2.9) has no solution for $\delta > \lambda_1/e$.

We ask next whether there are solutions to (2.9) for smaller values of δ. Clearly $\delta w(x)$, where w is the solution of (2.1), is a lower solution and we need only to find a larger upper solution to prove existence. Let us try an upper solution of the form $aw(x)$, which always satisfies the boundary inequality. We have $-\Delta(aw) = a$ which should exceed δe^{aw}. This is certainly accomplished if $a \geq \delta e^{a||w||}$ where $||w||$ is the maximum value of $w(x)$. Thus we need $\delta \leq a e^{-a||w||}$ where we are free to choose a. The largest δ range is obtained by taking $a = \frac{1}{||w||}$. Therefore $\frac{w}{||w||}$ is an upper solution whenever $\delta \le \frac{1}{e||w||}$. Note that $\frac{w}{||w||} > \delta w$ so that the upper solution exceeds the lower solution and there must therefore exist a solution of (2.9) for $\delta \le \frac{1}{e||w||}$ and the solution v satisfies

$$\delta w \le v \le \frac{w}{||w||} .$$

Combining with our previous result, we can guarantee a solution if $\delta \le \frac{1}{e||w||}$ and no solution for $\delta > \lambda_1/e$. A famous result of Pólya states that $\frac{1}{||w||} < \lambda_1$, so that our results are consistent but leave a gap in which the existence question has not been resolved.

The multiplicity results for (2.9) can be obtained for balls. The one-dimensional problem for a slab

$$- v'' = \delta e^{v} , \quad |x| < 1/2 ; \quad v(\pm \tfrac{1}{2}) = 0 ,$$

can be solved explicitly. We recall that δ is proportional to the square of the original thickness of the slab before nondimensionalization. It is clear that v is even, concave and has its maximum v_M at $x = 0$. Multiplying the differential equation in (2.11) by v', we can integrate once to obtain

$$v' = - [2\sigma(e^{v_M} - e^{v})]^{1/2} \quad 0 < x < \tfrac{1}{2} .$$

A further integration from x to $\frac{1}{2}$ yields

$$\int_0^{v} \frac{dz}{(e^{v_M} - e^{z})^{1/2}} = (2\delta)^{1/2} (\tfrac{1}{2} - x) ,$$

which, at $x = 0$, gives

$$\delta = 2I^2 ,$$

where

$$I(v_M) = \int_0^{v_M} \frac{dz}{(e^{v_M} - e^{z})^{1/2}} = 2e^{-v_M/2} \operatorname{arctanh} (1-e^{-v_M})^{1/2} .$$

We can plot v_M versus δ and interpret the figure.

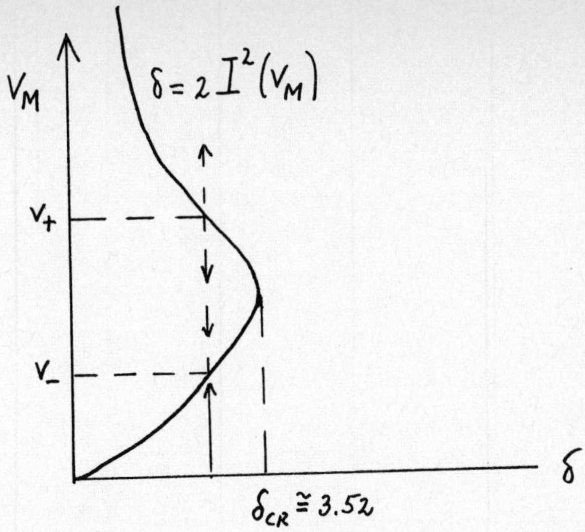

Fig. 3

For $\delta > \delta_{cr}$, there is no solution. For $\delta < \delta_{cr}$, there are two solutions, the smaller of which is stable. The arrows on the diagram indicate that the solution of the time-dependent problem for an initial value $v_0(x)$ below the larger steady-state solution $v_+(x)$ will tend as $t \to \infty$ to the smaller steady-state solution $v_-(x)$. If the initial value is above $v_+(x)$, the solution will become infinite as $t \to \infty$.

The analysis for a ball in R_3 is both more complicated and more interesting (see [16], for instance). The curve of δ vs u_M now has an infinite number of oscillations of decreasing amplitude. Again there is no solution for δ greater than some δ_{cr}, but there is a value of δ for which there exist infinitely many positive solutions! We shall not carry out the analysis here but turn instead to the time-dependent problem and ask what happens to the initial value problem if $\delta > \delta_{cr}$. We then have blow-up in finite time (see [15]). Let us prove this result for the general Gelfand problem.

$$\frac{\partial v}{\partial t} - \Delta v = \delta e^v , \quad x \in \Omega, \quad t > 0 ; \tag{2.12}$$
$$v(x,0) = 0 , \quad v(\partial\Omega,t) = 0 .$$

Here we have rescaled the time variable to include σ. Multiply (2.12) by $\phi_1(x)$, the positive fundamental eigenfunction of (2.10). Setting

$$E(t) = \int_\Omega \phi_1(x)\ v(x,t)dx \ ,$$

we obtain

$$\frac{dE}{dt} + \lambda_1 E = \delta \int_\Omega e^v\ \phi_1\ dx \ , \quad E(0) = 0 \ . \tag{2.13}$$

Jensen's inequality for convex functions yields (with $\int_\Omega \phi_1 dx = 1$)

$$\int_\Omega e^v\ \phi_1\ dx \geq \delta \exp\left(\int_\Omega v\phi_1\ dx\right) = \delta e^E \ ,$$

from which (2.13) becomes

$$\frac{dE}{dt} + \lambda_1 E \geq \delta e^E \quad \text{or} \quad \frac{dE}{dt} \geq \delta(e^E - \frac{\lambda_1}{\delta} E)$$

If $\frac{\lambda_1}{\delta} < e$ (that is, $\delta > \lambda_1/e$), then $e^E - \frac{\lambda_1}{\delta} E$ is positive for $E > 0$ and

$$\int_0^{E(t)} \frac{ds}{e^s - (\lambda_1 s/\delta)} \geq \delta t \ .$$

Since $K = \int_0^\infty \frac{ds}{e^s - (\lambda_1 s/\delta)}$ is finite, there is no solution beyond the time $t = K/\delta$. We conclude that there is blow-up before that time.

Next we turn to the catalyst particle with equal diffusivities. That case reduces to the scalar problem (2.4). Again we look first at the steady state problem (for m = 1):

$$- \Delta v = \delta(1 - \frac{\sigma}{\delta} v)\ e^{v/(1+\varepsilon v)} \ , \quad v(\partial\Omega) = 0 \ .$$

Existence of solutions for all δ follows from the boundedness of the nonlinearity. A sufficient condition for uniqueness is for

$$F(v) \doteq \left(\frac{1}{v} - \frac{\sigma}{\delta}\right) e^{v/(1+\varepsilon v)}$$

to be strictly decreasing. By the same analysis as in section I (following equation (1.26)), we have uniqueness if $\varepsilon > \frac{1}{4} - \frac{\sigma}{\delta}$. The inequality is automatically satisfied for all ε if $\delta < 4\sigma$. If $\varepsilon > \frac{1}{4}$, the inequality is satisfied for all σ and δ.

3. The stationary dead core

Consider an isothermal steady-state problem. The concentration then satisfies

$$- \Delta u = - \lambda u^m \ , \quad x \in \Omega \ ; \quad u = 1 \ , \quad x \in \partial\Omega \ , \tag{3.1}$$

where we have set $\lambda = 1/\mu$ (see equation (1.15)). It is, of course, understood that we are looking for solutions $u(x)$ such that $u(x) \geq 0$ in Ω. Such a solution also satisfies the equivalent extended problem

$$- \Delta u = - \lambda (u_+)^m \ , \quad x \in \Omega \ ; \quad u = 1 \ , \quad x \in \partial\Omega \ , \tag{3.2}$$

where $u_+ = \max(u,0)$. It is clear from the maximum principle that every solution of (3.2) satisfies (3.1) and is nonnegative in Ω. Also, by the maximum principle, $u(x) < 1$ in Ω. In the present section, we will be particularly interested in the possibility of a solution u vanishing in a subregion D of Ω. Such a region of zero concentration is known as a dead zone (or dead core). If $m \geq 1$, no such phenomenon can occur, whereas if λd^2 is sufficiently large - here d is a typical dimension - there will be a dead zone (see [11], [13], [5]).

The appearance of dead zones is easily established for one-dimensional problems. Let us begin by looking at the semi-infinite domain $x > 0$, leading to the boundary value problem

$$-u_{xx} = - \lambda u_+^m \ , \quad x > 0 \ ; \quad u(0) = 1 \ , \ u, \ u_x \to 0 \quad \text{as} \quad x \to \infty \ .$$

By multiplying by u_x, we can integrate to obtain (since u is a decreasing function of x)

$$u_x = - \left(\frac{2\lambda}{m+1}\right)^{1/2} u^{(m+1)/2} \ .$$

For $u > 0$, we can divide by $u^{(m+1)/2}$ to find

$$\left(\frac{m+1}{2}\right)^{1/2} \int_u^1 z^{-(m+1)/2} dz = \lambda^{1/2} x \ . \tag{3.3}$$

Writing

$$I = \left(\frac{m+1}{2}\right)^{1/2} \int_{0+}^1 z^{-(m+1)/2} dz \ ,$$

we see that when $I = \infty$ (that is when $m \geq 1$), then (3.3) yields a unique solution $u(x)$, positive for all x, with the property $\lim_{x\to\infty} u(x) = 0$. If, however, I is finite (that is, $m < 1$), then $u(x) > 0$ for $0 < x < \lambda^{-1/2} I$ and $u \equiv 0$ for $x \geq \lambda^{-1/2} I$. The solution obtained in this way, satisfies the differential equation for $x < \lambda^{-1/2} I$ and $x > \lambda^{-1/2} I$, is continuous, has a continuous derivative, and (if $m > 0$) has a continuous second derivative. Note that the value of I is given explicitly for $m < 1$ by $I = [2(m+1)]^{1/2}/(1-m)$.

We can say that there is complete penetration of the region $x > 0$ if $m \geq 1$, but only finite penetration if $m < 1$ with the penetration distance of order $\lambda^{-1/2}$. Thus, large values of λ coupled with $m < 1$ lead to the dead zones.

Turning next to a finite interval $0 < x < a$, we have the boundary value problem

$$- u_{xx} = - \lambda u_+^m , \quad 0 < x < a ; \quad u(0) = u(a) = 1 . \qquad (3.4)$$

If a dead core occurs it must be symmetric about $x = a/2$. It can be constructed by piecing together the dead zone solutions for the semi-infinite domains $x > 0$ and $x < a$. Therefore a dead core can occur only if $m < 1$ and $\lambda^{-1/2} I \le a/2$, that is, $\lambda a^2 \ge 4I^2$. The size of the dead core is then $a - 2I/\lambda^{1/2}$.

With these two simple examples available for guidance we now turn to the general problem (3.2). The existence of a solution of (3.2) follows from an application of the method of upper and lower solutions.

Definition. A function $\bar{u}(x)$ is said to be an upper solution of (3.2) if

$$- \Delta \bar{u} \ge - \lambda \bar{u}_+^m , \quad x \in \Omega ; \quad \bar{u} \ge 1 , \quad x \in \partial\Omega . \qquad (3.5)$$

Similarly, a lower solution $\underline{u}$ satisfies (3.5) with both inequalities reversed.

If there exist lower and upper solutions with $\underline{u}(x) \le \bar{u}(x)$ then (3.2) has a solution $u(x)$ satisfying $\underline{u} \le u \le \bar{u}$. Since $\underline{u} \equiv 0$ is a lower solution and $\bar{u} \equiv 1$ is an upper solution, there must exist a solution $u(x)$ of (3.2) satisfying $0 \le u(x) \le 1$. We have already seen that all solutions of (3.2) satisfy these inequalities. Further one can show that the solution is unique.

We have the following comparison theorems.

I. On a fixed domain Ω, consider problem (3.2) for two values of λ with $\lambda_2 \ge \lambda_1$. Then $u_2(x) \le u_1(x)$.

II. For fixed λ, consider (3.2) for two domains $\Omega_2 \supset \Omega_1$. Then $u_2(x) \le u_1(x)$ on Ω_1.

The proofs are simple. In I, $u_2(x)$ is a lower solution of (3.2) for $\lambda = \lambda_1$. In II, $u_2(x)$ is a lower solution of (3.2) formulated on Ω_1, since $u_2(\partial\Omega_1) \le 1$ by the maximum principle.

Now consider (3.2) on a bounded domain Ω, and let d be the thickness of the thinnest slab enclosing Ω. The slab solution (3.4) is then an upper solution to (3.2) on Ω. Therefore if the slab has no dead core, neither does Ω. In other words Ω has no dead core for $m \ge 1$, and has no dead core for $m < 1$ if $\lambda d^2 < 4I^2$.

Next we would like to show that, if $m < 1$, Ω must have a dead core for sufficiently large λ. We do this by showing first that a ball in R^n must have a dead core for sufficiently large λ. Assume a radially symmetric concentration $u(r)$ that must therefore satisfy

$$r^{1-n}(r^{n-1}u_r)_r = \lambda u^m , \quad 0 < r < a ; \quad u(a) = 1 . \tag{3.6}$$

We look for a solution of the form $u = \alpha r^p$. The Laplacian on the left side of (3.6) reduces the power by 2 so that, equating powers, we find that $p - 2 = mp$ or $p = 2/(1-m)$. Since $m < 1$, the function $\alpha r^{2/1-m}$ is smooth enough at $r = 0$ to have a classical Laplacian there. Substituting in (3.6), we find

$$\alpha^{1-m} = \frac{\lambda(1-m)^2}{2n(1-m) + 4m} . \tag{3.7}$$

It remains to satisfy the boundary condition at $r = a$. This will happen if $\alpha = a^{-2/(1-m)}$, that is, if $\lambda = \lambda^*$ where

$$\lambda^* a^2 = \frac{2n(1-m) + 4m}{(1-m)^2} \doteq P \tag{3.8}$$

To recapitulate: if $\lambda = \lambda^*$ and α is given by (3.7), we have found a solution $\alpha r^{2/(1-m)}$ of (3.6). Our solution vanishes at the center of the ball and is positive elsewhere. It thus represents a solution with a one-point dead core. By comparison Theorem I, the ball has a dead core for all λ greater that λ^* and no dead core for $\lambda < \lambda^*$.

We are now in a position to obtain dead core estimates for an arbitrary bounded domain Ω. Let r_i be the radius of the largest inscribed sphere and r_0 be the radius of the smallest circumscribed sphere. Then:

if $\lambda < P/r_0^2$, Ω does not have a dead core. (A)

if $\lambda \geq P/r_i^2$, Ω must have a dead core. (B)

if $\lambda \geq P/d^2$, where d is the distance from a point $x_0 \in \Omega$ to the boundary $\partial\Omega$, then x_0 belongs to the dead core. (C)

For a given x_0 in Ω, there is a threshold value of λ, say λ_c, such that x_0 belongs to the dead core if $\lambda \geq \lambda_c$ and does not if $\lambda < \lambda_c$. Of course, λ_c depends on x_0.

Another estimate can be obtained using the gradient bound

$$|\text{grad } u|^2 \leq 2\lambda \int_M^{u(x)} t^m dt \quad (M: \min_{x\in\Omega} u(x)), \tag{3.9}$$

derived in [20] for any domain Ω whose boundary has nonnegative average curvature. Let A be a point on the boundary of the dead core and let Q be the nearest point on $\partial\Omega$. Join A and Q by a straight line on which a coordinate r has been introduced with origin at A. Then, along this line, we obtain

$$\left(\frac{du}{dr}\right)^2 \le |\mathrm{grad}\, u|^2 \le 2\lambda \int_0^u t^m\, dt = \frac{2\lambda}{m+1} u^{m+1}$$

or

$$\frac{du}{dr} \le [2\lambda/(m+1)]^{1/2} u^{(m+1)/2} ,$$

which, on further integration from A to Q, becomes

$$I \le \lambda^{1/2} |AQ| \qquad (I \text{ given after } (3.3))$$

Thus, a point x_0 cannot be in the dead core unless its distance d from the boundary exceeds $I\lambda^{-1/2}$. Combining this result with c), we find that the dead core $D(\lambda)$ satisfies

$$S(\sqrt{P/\lambda}) \subset D(\lambda) \subset S(I/\sqrt{\lambda}) , \qquad (3.10)$$

where $S(d)$ is the set of points whose distance from the boundary is at least d. More delicate estimates can be found in [13] but it is clear from (3.10) that, for large λ, the dead core is at a distance $0(\lambda^{-1/2})$ from the boundary of Ω.

Let $u(x)$ be the solution of (3.2). In dealing with sets in R^n, we use the suggestive terms volume and area instead of the more precise n-measure and (n-1)-measure. The distribution function of u is the volume $\mu(t)$ of the set in R^n where $u(x) \le t$. Clearly, $\mu(t)$ is defined for $M \le t \le 1$ where M is the minimum of $u(x)$. We also note that $\mu(1) = |\Omega|$, $\mu(M) = 0$ unless there is a dead core D in which case $\mu(0) = |D|$. It can be seen from the differential equation that the volume of the set $u(x) = t$ is zero for $t > 0$. Thus, $u(t)$ is strictly increasing and its inverse $t(\mu)$ is defined and strictly increasing for $\mu \ge |D|$. If $|D| > 0$, we extend t to be identically zero for $0 \le \mu < |D|$. Note that $t(0) = M$ and $t(|\Omega|) = 1$. the function $t(\mu)$ is known as the increasing rearrangement of $u(x)$.

Under mild conditions, an easy calculation shows that

$$\frac{d\mu}{dt} = \int_{u=t} \frac{ds}{|\mathrm{grad}\, u|} \qquad t > M , \qquad (3.11)$$

where ds is an element of area on the surface $u = t$, whose total surface area will be denoted by $s(t)$. Setting

$$E(t) = \lambda \int_{u \le t} u^m(x)\, dx ,$$

we find from (3.2) that

$$E(t) = \int_{u=t} |\mathrm{grad}\ u|\, ds = \text{flux of } u \text{ through } u = t .$$

Combining this equation with (3.11), we obtain by using Schwarz's inequality

$$E(t) \frac{d\mu}{dt} = \int_{u=t} |\mathrm{grad}\ u|\, ds \int_{u=t} \frac{ds}{|\mathrm{grad}\ u|} \ge s^2(t) , \quad t > M .$$

We can rewrite this inequality in terms of μ as independent variable and then apply the classical isoperimetric inequality to obtain

$$\bar{E}(\mu) \ge q(\mu) \frac{dt}{d\mu} \tag{3.12}$$

where $\bar{E}(\mu) := E(t(\mu))$ and $q(\mu)$ is the square of the surface area of a ball of volume μ. In view of the extended definition of $t(\mu)$ it is clear that (3.12) holds for $\mu \ge 0$. We also note from the definition of $\dot{E}(t)$ and (3.11) that

$$\frac{d\bar{E}}{d\mu} = \frac{dE}{dt}\frac{dt}{d\mu} = \lambda t^m \frac{d\mu}{dt}\frac{dt}{d\mu} = \lambda t^m(\mu)$$

so that

$$\bar{E}(\mu) = \lambda \int_0^\mu t^m(\nu)\, d\nu . \tag{3.13}$$

The solution $u^*(x)$ of (3.2) for a ball Ω^* of the same volume as Ω is radially symmetric. As a consequence, (3.12) becomes an equality and

$$\bar{E}^*(\mu) = \lambda \int_0^\mu [t^*(\nu)]^m\, d\nu = q(\mu) \frac{dt^*}{d\mu} \tag{3.14}$$

where t^* is the increasing rearrangement of u^* and $\bar{E}^*$ is the flux of u^* through the spherical surface (level set for u^*) enclosing the volume μ.

Combining (3.12), (3.13), (3.14), we find

$$q(\mu) \left[\frac{dt}{d\mu} - \frac{dt^*}{d\mu}\right] \le \bar{E}(\mu) - \bar{E}^*(\mu) = \lambda \int_0^\mu \{[t(\nu)]^m - [t^*(\nu)]^m\}\, d\nu ,$$

$$\mu \ge 0 . \tag{3.15}$$

From (3.15), we will prove

Theorem. Suppose that $u(x)$ is a solution of (3.2) in a bounded domain Ω of R^n. Let $M := \min u$ and M^* the minimum of the solution $u^*(x)$ of the corresponding problem for a ball Ω^* of the same volume as Ω. Then one has

$$M \geq M^* \tag{3.16}$$

so that Ω does not have a dead core if Ω^* has no dead core. Moreover, the volumes of the dead cores satisfy $|D| \leq |D^*|$.

Proof. Let $t(\mu)$ and $t^*(\mu)$ be the increasing rearrangements of $u(x)$ and $u^*(x)$, respectively. We shall prove that there cannot exist an interval $[0,b]$ on which $t^* \geq t$ and $t^*(a) > t(a)$ for some a, $0 \leq a < b$. Letting b be the right end-point of the maximal interval of this type, we must have $t^*(b) = t(b)$ since $t^*(|\Omega|) = t(|\Omega|)$. If such an interval $[0,b]$ existed, then (3.16) shows that

$$(t-t^*)' \leq 0 \quad \text{on} \quad 0 < \mu < b$$

and, integrating from a to b, we find

$$t^*(a) \leq t(a)$$

contradicting our assumption.

An immediate conclusion is that $t^*(0) \leq t(0)$ (otherwise, setting $a = 0$, there would exist an interval $[0,b]$ of the type excluded). This proves (3.16) and, therefore Ω can have a dead core only if Ω^* has one. It also follows that $|D| \leq |D^*|$ for, otherwise, we would have $|D^*| < |D|$ and, setting $a = |D|$, there would exist an interval $[0,b]$ of the type excluded by our previous argument. This establishes the theorem.

The theorem leads to an improvement in the previous result (A): Let $0 \leq m < 1$. Then Ω has no dead core if

$$\lambda < P/R^2 ,$$

where R is the radius of an n-ball of volume $|\Omega|$. (D)

4. The formation of the dead core

The time-dependent version of (3.2) is the scalar parabolic problem

$$u_t - \Delta u = - \lambda u_+^m , \quad x \in \Omega , \quad t > 0 ; \tag{4.1}$$

$$u(x,0) = 1 \quad \text{for} \quad x \in \Omega \tag{4.2}$$

$$u(\partial\Omega,t) = 1 \quad \text{for} \quad t > 0 . \tag{4.3}$$

The problem (4.1), (4.2), (4.3) describes the evolution of the concentration $u(x,t)$ in an isothermal reaction. Equation (4.1) is obtained from (1.15) by setting $v = 0$, $\lambda = 1/\mu$ and by incorporating μ in a rescaled time variable. We are interested in how the solution $u(x,t)$ approaches $U(x)$, the solution of the steady state problem (3.2). Because $u(x,0) > U(x)$ in Ω, it can be shown that $u(x,t)$ decreases monotonically in time to $U(x)$. It can also be shown (see [4]) that $u(x,t)$ is strictly decreasing in t as long as $u(x,t) > 0$. Thus, if $m \geq 1$, the time-dependent solution does not reach the steady state (which is, of course, positive) in finite time. If $m < 1$ and if λ is sufficiently large, we know that the stationary problem has a dead core. Does the corresponding solution of the parabolic problem then vanish in finite time? For the spatially homogeneous case (which satisfies (4.1), (4.2) and, instead of (4.3), the no flux condition $\partial u/\partial \nu = 0$), we know that the solution $p(t)$ is identically zero after some finite time (with $m < 1$, of course). In our case, however, the boundary condition (4.3) supplies reactant continuously and it is not clear whether the concentration can be zero in finite time. If this is possible, it can happen only for points x in the dead core D of the stationary problem (with the same λ). Which points x_0 in D have this property and at what time does $u(x_0,t)$ vanish? Our principal result is contained in the following.

Theorem. Let $x_0 \in \Omega$ and $m < 1$ be fixed. As we have seen in section 3, x_0 will belong to the dead core $D(\lambda)$ of the stationary problem if $\lambda \geq \lambda_c$, where λ_c is the threshold value for x_0. We now claim that x_0 will also belong to the dead core of the time-dependent problem for $t \geq \dfrac{1}{(1-m)(\lambda-\lambda_c)}$.

Proof. We shall construct an upper solution $z(x,t)$ of (4.1), (4.2), (4.3) for an interval $(0,T_0)$ with $z(x_0,T_0) = 0$. Since $u(x,t) \leq z(x,t)$ on $(0,T_0)$ it follows that $u(x_0,T_0) = 0$ and, by the monotonicity of u, $u(x_0,t) \equiv 0$ for $t \geq T_0$. We now exhibit $z(x,t)$ by setting

$$z(x,t) = U(x,\lambda_c) + p(\alpha t)$$

where α is a parameter yet to be determined, $U(x,\lambda_c)$ is the steady state solution for $\lambda = \lambda_c$ and $p(t)$ is the solution of the spatially homogeneous problem

$$p'(t) = - p_+^m \, , \quad p(0) = 1 \, , \tag{4.4}$$

whose solution is given explicitly by

$$p(t) = H\left(\frac{1}{m-1} - t\right) [1 - (1-m)t]^{1/(1-m)} \, ,$$

H being the Heaviside function. Note that $p(t)$ vanishes identically for $t \geq 1/(1-m)$ and $p(\alpha t)$ for $t \geq 1/\alpha(1-m)$. We see that $z(x,0) \geq 1$, $z(\partial\Omega,t) \geq 1$ and $z\left(x_0, \frac{1}{\alpha(1-m)}\right) = 0$ so that we only need to establish the differential inequality

$$z_t - \Delta z \geq -\lambda z^m, \text{ for } t < 1/\alpha(1-m). \tag{4.5}$$

A straightforward calculation gives

$$\begin{aligned} z_t - \Delta z &= \alpha p'(\alpha t) - \Delta U = -\alpha p^m - \lambda_c U^m \\ &\geq -(\alpha+\lambda_c)\,[\max\,(p\ ,U\ )]^m \\ &\geq -(\alpha+\lambda_c)(p\ +U\ )^m = -(\alpha+\lambda_c)\ z^m\ . \end{aligned}$$

By choosing $\alpha = \lambda - \lambda_c$, we obtain the desired inequality (4.5) for z.

Remark. Since λ_c is not known explicitly, we cannot use the inequality $t \geq \frac{1}{(1-m)(\lambda-\lambda_c)}$ directly. However, we have bounds for λ_c in section 3. In fact condition (C) shows that $\lambda_c \leq P/d^2$, where d is the distance from x_0 to the boundary $\partial\Omega$ and P is defined in (3.8). Therefore if $\lambda > P/d^2$ then x_0 belongs to the dead core of the parabolic problem for

$$t \geq \frac{1}{(1-m)[\lambda-(P/d^2)]}\ .$$

Since $\max\limits_{x_0 \in \Omega} d = r_i$, we can estimate the time τ of onset of the dead core (that is, the time at which $u(x,\tau)$ is first zero some place in Ω). Let $\lambda > P/r_i^2$, then the time of onset τ satisfies

$$\tau \leq \frac{1}{(1-m)[\lambda-(P/r_i^2)]}\ .$$

We can restate our result in a somewhat different form. Let λ be large enough so that the stationary problem has a dead core D. If x_0 is an interior point of D, then x_0 belongs to the dead core of the parabolic problem for t sufficiently large. Obviously points in D near ∂D will take longer to join the parabolic dead core. Points on ∂D do not belong to the parabolic core for any finite time. To prove this, let $D(t)$ be the parabolic dead core at time t which we may assume to consist of only one component. We shall show that $\text{dist}(\partial D(t),\ D(t-\Delta t)) > 0$. Suppose this assertion is false. Then there exists a point $x_0 \in \partial D(t_0-\Delta t) \cap \partial D(t_0)$. Let R_μ be the set of points (x,t) such that $u(x,t) \leq \mu$, $t \leq t_0$, and set

$$T_\mu = \overline{\partial R_\mu - \{(x,t) : t = t_0\}} \ .$$

On T_μ, we have $u(x,t) = \mu > 0$ and therefore there exists $\delta > 0$ such that $v(x,t) \doteq u(x,t-\Delta t) \geq \mu + \delta$. Hence the set $K \doteq \{(x,t) : v(x,t) = \mu + \frac{\delta}{2}, \ t < t_0\}$ is strictly contained in R_μ, and the distance between K and T_μ on lines t = constant is larger than some positive number ε. Take any unit vector e in R^n such that $x_0 + \varepsilon e$ is <u>not</u> in $D(t_0)$. Define $v_\varepsilon(x,t) = v(x-\varepsilon e,t)$. Clearly, v_ε satisfies (4.1) in R_μ and $v_\varepsilon \geq \mu + \delta/2$ on T_μ. Therefore v_ε is an upper solution (≤ 1) to (4.1), (4.2), (4.3) on R_μ and, consequently,

$$v_\varepsilon(x,t) \geq u(x,t) \quad \text{in} \quad R_\mu \ .$$

However, we also have

$$v_\varepsilon(x_0+\varepsilon e,t_0) = u(x_0,t_0-\Delta t) = 0 < u(x_0+\varepsilon e,t_0) \ ,$$

thereby contradicting the previous inequality.

We have therefore proved the distance between the boundaries of the dead core at times $t - \Delta t$ and t is positive. It follows that a boundary point of the steady-state dead core cannot belong to $D(t)$ for any finite time. For other results concerning the formation of the dead core, see [4] and [12].

5. Gas-solid reactions

In this section, we shall consider a new type of problem involving two reacting substances, one a diffusing gas, the other an immobile solid. As the gas diffuses through the porous solid, a reaction takes place whose rate depends on the concentrations of both the gas and the solid. Various processes that can be modeled in this way are described in [14] and [23]; they include ore reduction, retorting of oil shale, and catalyst regeneration, and, by changing the direction of the solid reaction, catalyst deactivation. To fix ideas, we shall consider only the simplest case of the combustion of a porous solid as it reacts with a diffusing gas. The reaction is assumed to be irreversible and proceed <u>isothermally</u>. Under the assumption of equimolar counterdiffusion and constant effective diffusivity and porosity, the nondimensional equations of mass conservation of the two phases become

$$\varepsilon \frac{\partial C}{\partial t} - \Delta C = -\phi^2 f(C,S) \tag{5.1}$$

$$\frac{\partial S}{\partial t} = -f(C,S) \tag{5.2}$$

which hold for $t > 0$ in the domain Ω occupied by the solid. The associated initial conditions are chosen as

$$C(x,0) = 0 \;, \qquad S(x,0) = 1 \tag{5.3a,b}$$

and the boundary condition as

$$C(\partial\Omega,t) = 1 \;, \qquad t > 0 \;. \tag{5.4}$$

Here $C(x,t)$ is the nondimensional gas concentration referred to its constant ambient value and $S(x,t)$ is the nondimensional solid concentration referred to its constant initial value. The nondimensional porosity ε is usually of the order 10^{-1} or 10^{-2} while the Thiele's modulus ϕ^2, which measures the relative strength of reaction to diffusion, is in the range of 1 to 100 in the problems of interest to us. In nondimensionalizing the length scale, the volume $|\Omega|$ of Ω has been made equal to 1.

The function $f(C,S)$ will be taken of the form CS^m where $m \geq 0$. This type of reaction rate occurs frequently in practice and is essential for some, but not all, of our calculations.

The range of values of ϕ^2 is one in which both reaction and diffusion are significant. The limiting cases where diffusion or reaction dominate are easily analyzed. If diffusion dominates, both ϕ^2 and ε are small and C is nearly 1 throughout Ω and S decays everywhere in Ω according to the law $\partial S/\partial t = -S^m$. For $m \geq 1$, $S > 0$ for all x,t but $\lim_{t\to\infty} S(t) = 0$. For $m < 1$, S will be identically zero after $t = 1/(1-m)$. If reaction dominates, the reaction first takes place near the boundary where C is largest and the solid there will be almost fully converted before the gas can penetrate the solid to any appreciable depth. Thus, we will have a narrow reaction zone moving inward through the solid - the so-called "shrinking core" model. We are interested in the general problem in which neither limiting analysis is appropriate.

We now turn our attention to the problem with the particular form of f mentioned earlier. We are seeking functions $C(x,t) \geq 0$ and $S(x,t) \geq 0$ satisfying

$$\begin{cases} \varepsilon \dfrac{\partial C}{\partial t} - \Delta C = -\phi^2 CS^m & , \quad x \in \Omega, \quad t > 0 \\[2mm] \dfrac{\partial S}{\partial t} = -CS^m & , \quad x \in \Omega, \quad t > 0 \\[2mm] C(x,0) = 0, \quad S(x,0) = 1, \quad C(\partial\Omega,t) = 1 \quad \text{for} \quad t > 0 \;. \end{cases} \tag{5.5}$$

We are interested in studying the validity of a popular approximation known as the <u>pseudo-steady-state approximation</u>, in which ε is set equal to zero in the first equation in (5.5); we shall also obtain estimates for the <u>conversion</u> $\gamma(t)$, the amount of solid that has been converted to products by time t.

We begin by making some simple observations with respect to (5.5). Clearly $S(x,t)$ is decreasing in time; if $S(x_0,T) = 0$ then $S(x_0,t) \equiv 0$ for $t \geq T$. We can also show that $C(x,t)$ is increasing in time (because of our choice of initial conditions). To show this, think of $S(x,t)$ as known so that the first equation in (5.5) is a scalar parabolic equation for C with initial condition $C(x,0) = 0$ and boundary condition $C(\partial\Omega,t) = 1$. The function $v(x,t) = C(x,t+\Delta t)$ satisfies the same equation with $S^m(x,t+\Delta t)$ replacing $S^m(x,t)$. But by the monotonicity of S, this means that

$$\varepsilon \frac{\partial v}{\partial t} - \Delta v \geq -\phi^2 v\, S^m(x,t)$$

and clearly $v(x,0) \geq 0$, $v(\partial\Omega,t) = 1$. Thus, $v(x,t)$ is an upper solution for the scalar problem for C and hence $C(x,t+\Delta t) \geq C(x,t)$ and $C(x,t)$ is increasing in time. This also means that $\partial C/\partial t \geq 0$. It is also clear that $\overline{C} \equiv 1$ is an upper solution so we conclude that $C(x,t)$ is increasing in t and $C(x,t) \leq 1$. The steady state for (5.5) is $S \equiv 0$ and $C \equiv 1$; we shall show later that the solution (C,S) of (5.5) actually tends to $(1,0)$ as $t \to \infty$.

If the diffusivity is large then both ϕ^2 and ε are small (the equation for C was obtained by dividing the original dimensional equation by the diffusivity). Thus, C is essentially equal to 1 throughout Ω for all t. This yields the largest possible conversion of the solid. Setting $C = 1$ in the equation for S, we again obtain the familiar

$$\frac{dS}{dt} = -S^m, \qquad t > 0 ; \qquad S(0) = 1 . \tag{5.6}$$

The amount of solid converted in this case of "fast diffusion" is $A = 1 - S$ where $A(t)$ has the explicit form

$$A(t) = \begin{cases} 1 - [1 + t(m-1)]^{-1/(m-1)} & m > 1 , \\ 1 - e^{-t} & m = 1 , \\ 1 - H(\frac{1}{m-1} - t)[1 - (1-m)t]^{1/(1-m)} & m < 1 . \end{cases} \tag{5.7}$$

In all 3 cases, $A(t)$ increases monotonically to 1 as $t \to \infty$. If $m < 1$, A is identically 1 after the finite time $t = 1/(1-m)$. Figure 4 illustrates $A(t)$ for different values of m.

Returning to the general case, we have, for $S > 0$

$$-\frac{1}{S^m}\frac{\partial S}{\partial t} = C$$

from which follows that

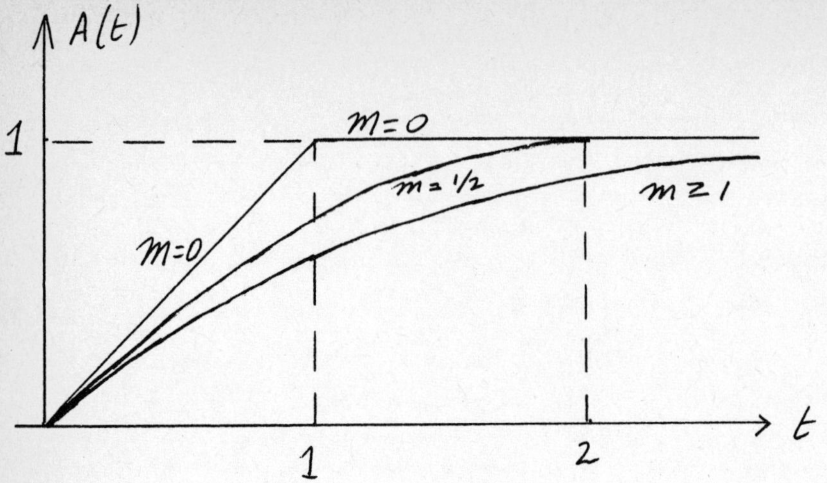

Fig. 4

$$\int_S^1 \sigma^{-m}\, d\sigma = \psi(x,t) \ , \qquad (5.8)$$

where the cumulative (gas) concentration ψ is defined as

$$\psi(x,t) = \int_0^t C(x,\tau)\, d\tau \ .$$

The formula for (5.6) corresponding to (5.8) is

$$\int_S^1 \sigma^{-m}\, d\sigma = t \ .$$

Thus, the solution of 5.8 is

$$1 - S(x,t) = A(\psi(x,t))$$

where A is given by (5.7). Note that if $m \geq 1$, A is always less than one so that $S(x,t) > 0$ for all x,t. If, however, $m < 1$, $S(x,t) = 0$ when $\psi(x,t) \geq 1/(1-m)$. Since ψ is largest on $\partial\Omega$ it is clear that S is smallest on $\partial\Omega$ and conversion of the solid begins on $\partial\Omega$ and then moves inward in time. The solid core therefore shrinks as time progresses - we can regard this problem as a moving boundary problem. When

$$\min_{x\in\Omega} \psi(x,t) = 1/(1-m) \ , \qquad (5.9)$$

the solid has been fully converted. Equation (5.9) characterizes the time t_1 for full conversion.

We can now reduce (5.5) to a scalar problem for ψ. Integrate the equation for C with respect to time to obtain

$$\begin{cases} \varepsilon \frac{\partial \psi}{\partial t} - \Delta \psi = -\phi^2 A(\psi) , & x \in \Omega , \quad t > 0 , \\ \psi(x,0) = 0 , \quad \psi(\partial\Omega,t) = t & \text{for} \quad t > 0 . \end{cases} \tag{5.10}$$

Clearly $0 \leq \psi \leq t$, and both ψ and $\partial\psi/\partial t$ increase in time. Problem (5.10) has one and only one solution. Let us set

$$\psi(x,t) = t - \eta(x,t) \tag{5.11}$$

to obtain

$$\varepsilon \frac{\partial \eta}{\partial t} - \Delta \eta = \varepsilon + \phi^2 A(t-\eta) , \quad \eta(x,0) = \eta(\partial\Omega,t) = 0 . \tag{5.12}$$

With $w(x)$ the solution of the Poisson problem (2.1), we can show that $(\varepsilon+\phi^2)w$ is an upper solution of (5.12) so that

$$t - (\varepsilon+\phi^2)w(x) \leq \psi(x,t) \leq t . \tag{5.13}$$

Since $\psi = \int_0^t C(x,\tau)\, d\tau$ and C is increasing in t, it follows from (5.13) that $C(x,t) \to 1$ as $t \to \infty$. The second equation in (5.5) then shows that $S \to 0$ as $t \to \infty$.

Next, we introduce the <u>pseudo-steady-state</u> problem by setting $\varepsilon = 0$ in (5.5). This yields the boundary value problem

$$\begin{cases} -\Delta\tilde{C} = -\phi^2 \tilde{C}\tilde{S}^m , & x \in \Omega \\ \frac{\partial \tilde{S}}{\partial t} = -\tilde{C}\tilde{S}^m , & x \in \Omega , \quad t > 0 \\ \tilde{S}(x,0) = 1, & \tilde{C}(\partial\Omega,t) = 1 \end{cases} \tag{5.14}$$

The gas concentration $\tilde{C}$ still depends on time through $\tilde{S}$ but, since the equation for the gas concentration is now elliptic, we cannot impose any initial condition. Indeed $\tilde{C}(x,0)$ satisfies

$$-\Delta\tilde{C} = -\phi^2\tilde{C} , \quad x \in \Omega , \quad \tilde{C}(\partial\Omega,0) = 1 , \tag{5.15}$$

whose unique solution $e(x)$ has the properties

$$0 < e(x) < 1 .$$

Again, we see that $\tilde{S}$ decreases in t and therefore $\tilde{C}$ increases. By the maximum principle, it is clear that $\tilde{C} \leq 1$. We also have $\tilde{C}(x,t) \geq e(x)$ so that

$$e(x) \leq \tilde{C} \leq 1 .$$

The cumulative gas concentration is given by

$$\tilde{\psi}(x,t) = \int_0^t \tilde{C}(x,\tau)\, d\tau$$

and $\tilde{\psi}$ satisfies the elliptic problem

$$- \Delta\tilde{\psi} = - \phi^2 A(\tilde{\psi}) \ , \quad \tilde{\psi}(\partial\Omega,t) = t \ , \qquad (5.16)$$

where the dependence of $\tilde{\psi}$ on t arises solely from the boundary condition. Problem (5.1b) has one and only one solution $\tilde{\psi}$ which is easily seen to satisfy

$$t - \phi^2 w(x) \le \tilde{\psi} \le t \ ,$$

where $w(x)$ is the solution of (2.1). More interestingly $\tilde{\psi}$ is an upper solution to (5.10). Indeed we have $\frac{\partial\tilde{\psi}}{\partial t} = \tilde{C}$ so that $0 \le \frac{\partial\tilde{\psi}}{\partial t} \le 1$ and

$$\varepsilon \frac{\partial\tilde{\psi}}{\partial t} - \Delta\tilde{\psi} \ge - \phi^2 A(\tilde{\psi}) \ , \quad \tilde{\psi}(x,0) = 0 \ , \quad \tilde{\psi}(\partial\Omega,t) = t \ ,$$

and therefore

$$\psi(x,t) \le \tilde{\psi}(x,t) \ .$$

We also observe that

$$z(x,t) = \tilde{\psi} - \varepsilon w(x)$$

is a lower solution to (5.10). We collect some of these inequalities:

$$\tilde{\psi}(x,t) - \varepsilon w(x) \le \psi(x,t) \le \tilde{\psi}(x,t) \le t \ . \qquad (5.17)$$

Note that we have shown that

$$\sup_{(x,t)} | \ \psi(x,t) - \tilde{\psi}(x,t) \ | \le \varepsilon \ ||w|| \ ,$$

where $||w|| = \max_{x\varepsilon\Omega} w(x)$. Thus $\tilde{\psi}$ gives a uniformly $0(\varepsilon)$ approximation to ψ. This justifies the engineering use of the approximation.

Next we shall study the time required to convert the solid to products. For the case of "fast diffusion", $C \equiv 1$ and conditions are spatially homogeneous so that the total amount converted up to time t is just $A(t)$. For the exact problem (5.5) or (5.10) the amount converted up to time t is

$$\gamma(t) = \int_{\Omega} A(\psi(x,t))dx \ , \qquad (5.18)$$

where we recall that $|\Omega| = 1$. In the pseudo-steady-state approximation, the conversion is given by

$$\tilde{\gamma}(t) = \int_{\Omega} A(\tilde{\psi}(x,t))dx \ .$$

In view of (5.17) we have

$$\omega(t) \le \gamma(t) \le \tilde{\gamma}(t) \le A(t)$$

where

$$\omega(t) = \int_{\Omega} A(\tilde{\psi}(x,t) - \varepsilon w(x))dx$$

The times at which the fraction p of the solid is converted satisfy the inequalities

$$A^{-1}(p) \le \tilde{\gamma}^{-1}(p) \le \gamma^{-1}(p) \le \omega^{-1}(p) .$$

For the case in which full conversion occurs $(m<1)$, we can find an explicit expression for the time τ_1 at which the solid is fully converted in the pseudo-steady-state approximation..

For the solid to be fully converted, we need

$$A(\tilde{\psi}(x,t)) = 1 \quad \text{for all} \quad x ,$$

or

$$\tilde{\psi} \ge 1/(1-m) \quad \text{for all} \quad x .$$

The time τ_1 at which this inequality is first satisfied for all x is characterized by

$$\underset{x\varepsilon\Omega}{\text{Min}}\ \tilde{\psi}(x,\tau_1) = 1/(1-m) . \tag{5.19}$$

When $t \ge \tau_1$, $\tilde{\psi}$ satisfies

$$-\Delta\tilde{\psi} = -\phi^2 , \quad x \in \Omega ; \quad \tilde{\psi}(\partial\Omega,t) = t$$

so that

$$\tilde{\psi} = t - \phi^2 w(x) , \quad t \ge \tau_1 ,$$

and

$$\tilde{\psi}(x,\tau_1) = \tau_1 - \phi^2 w(x) ,$$

$$\underset{x\varepsilon\Omega}{\text{Min}}\ \tilde{\psi}(x,\tau_1) = \tau_1 - \phi^2 \,||w|| . \tag{5.20}$$

Setting this expression equal to (5.19), we have

$$\tau_1 = \frac{1}{1-m} + \phi^2 \,||w|| . \tag{5.21}$$

The quantity $||w||$ is known for some simple domains and there are various good estimates available in other cases (see [3]). With t_1 the time for full conversion in the exact problem, we find from (5.17)

$$\tau_1 \le t_1 \le \tau_1 + \varepsilon \,||w|| \tag{5.22}$$

where τ_1 is given by (5.21).

For a slab of thickness a, $w(x) = \frac{x(a-x)}{2}$, and $||w|| = a^2/8$. Thus

$$\frac{1}{1-m} + \phi^2 \frac{a^2}{8} \le t_1 \le (\phi^2+\varepsilon) \frac{a^2}{8} + \frac{1}{1-m} ,$$

giving a maximum error of $\varepsilon a^2/8$.

By the same rearrangement methods used in chapter 3 and in [27], it is also possible to show that, among all domains of equal volume, conversion is slowest for the ball.

Bibliography

[1] ARIS, R. The mathematical theory of diffusion and reaction in permeable catalysts, Clarendon Press, Oxford, 1975.

[2] ARIS, R. On stability criteria of chemical reaction engineering, Chemical Engineering Science, 24 (1969), 149-169.

[3] BANDLE, C. Isoperimetric inequalities and their applications, Pitman, London (1980).

[4] BANDLE, C. and STAKGOLD, I., The formation of the dead core in parabolic reaction-diffusion equations, Trans. Amer. Math. Soc., 286 (1984), 275-293.

[5] BANDLE, C., SPERB, R. P. and STAKGOLD, I., Diffusion-reaction with monotone kinetics, Nonlinear Analysis, 8 (1984), 321-333.

[6] BODDINGTON, T., GRAY, P. and WAKE, G. C., Criteria for thermal explosions with and without reactant consumption, Proc. R. Soc. Lond. A., 357 (1977), 403-422.

[7] BUCKMASTER, J. D. and LUDFORD, G. S. S., Lectures on mathematical combustion, SIAM, 1983.

[8] BURNELL, J. G., LACEY, A.A. and WAKE, G. C., Steady states of the reaction-diffusion equations; Parts I and II, J. Austral. Math. Soc. B., 24 (1983), 374-416.

[9] CHANDRA, J. and DAVIS, P. WM., Rigorous bounds and relations among spatial and temporal approximations in the theory of combustion, Combustion Science and Technology, 23 (1980), 153-162.

[10] CHANDRA, J. and DAVIS, P. WM., Comparison theorems for systems of reaction-diffusion equations, in Applied Nonlinear Analysis, Academic Press, 1979.

[11] DIAZ, J. I. and HERNANDEZ, J., On the existence of a free boundary for a class of reaction diffusion systems, SIAM J. Math. Anal., 15 (1984), 670-685.

[12] DIAZ, J. I. and HERNANDEZ, J., Some results on the existence of free boundaries for parabolic reaction-diffusion systems, Proc.

Fifth Internat. Conf. on Trends in Theory and Practice of Nonlinear Differential Equations, Dekker, New York, 1984.

[13] FRIEDMAN, A. and PHILLIPS, D., The free boundary of a semilinear elliptic equation, Trans. Amer. Math. Soc., 282 (1984), 153-182.

[14] FROMENT, G. F. and BISCHOFF, K. B., Chemical reactor design and analysis, Wiley & Sons, New York, NY (1979).

[15] FUJITA, H., On the nonlinear equations $\Delta u + e^u = 0$ and $\partial v/\partial t = \Delta v + e^v$, Bull. Amer. Math. Soc., 75 (1969), 132-135.

[16] JOSEPH, D. D. and LUNDGREN, T. S., Quasilinear Dirichlet problems driven by positive sources, Arch. Ration. Mech. Anal., 49 (1973), 241.

[17] KASSOY, D. R. and LIÑAN, A., The influence of reactant consumption on the critical conditions for homogeneous thermal explosions Q. Jour. Mech. Appl. Math., 31 (1978), 99-112.

[18] LACEY, A. A., Critical behaviour of homogeneous reacting systems with large activation energy, Int. J. Eng. Science, 21 (1983), 501-515.

[19] PAO, C. V., Asymptotic stability of reaction-diffusion systems in chemical reactor and combustion theory, J. Math. Anal. Appl., 82 (1981), 503-526.

[20] PAYNE, L. E. and STAKGOLD, I., Nonlinear problems in nuclear reactor analysis in Lecture Notes in Math., 322, Springer, 1973.

[21] POORE, A. B., A model equation arising from chemical reactor theory, Arch. Rat. Mech. Anal., 52 (1973), 358-388.

[22] SATTINGER, D. H., Topics in stability and bifurcation theory, Lecture Notes in Mathematics 309, Springer, Berlin (1973).

[23] SOHN, H. Y. and SZEKELY, J., A structural model for gas-solid reactions with a moving boundary, Chem. Eng. Sci., 27 (1972), 763-778.

[24] STAKGOLD, I., Gas-solid reactions, in Dynamical Systems II, A. R. Bednarek and L. Cesari, eds., Academic Press, New York, 1982.

[25] STAKGOLD, I., BISCHOFF, K. B., and GOKHALE, V., Validity of the pseudo-steady-state approximation, Int. J. of Eng. Sci., 21 (1983), 537-542.

[26] STAKGOLD, I. and McNABB, A., Conversion estimates for gas-solid reactions, Math. Modelling, 5 (1984), 325-330.

[27] TALENTI, G., Elliptic equations and rearrangements, Annali. Scu. norm. sup. Pisa, Ser 4, 3 (1976), 697.

[28] VASQUEZ, J. L., A strong maximum principle for some quasilinear elliptic equations, Appl. Math. and Optimization, 12 (1984), 191-202.

REARRANGEMENTS OF FUNCTIONS AND PARTIAL DIFFERENTIAL EQUATIONS

Giorgio Talenti

§ 1. Introduction.

The following BASIC code:

```
10  PRINT "how many words"; : INPUT N
20  DIM V$(N) : V$(0) = "A"
25  FOR I = 1 TO N
30  READ WRD$
35  J = I - 1
40  IF V$(J) < = WRD$ THEN 55
45  V$(J + 1) = V$(J)
50  J = J - 1 : GOTO 40
55  V$(J + 1) = WRD$
60  NEXT I
70  FOR I = 1 TO N : PRINT I, V$(I) : NEXT I
80  DATA ... ,
               , ...
81  DATA ... ,
               , ...
.
.
.
100 END
```

describes nothing but a customary job - for instance, what every teacher does for arranging in alphabetical order the homeworks of his students. The code in question produces the alphabetical list of the first N words of a text. The output may look freakish or boring, and surely should not be used for enjoying poetic properties. However, it might prove suitable for analyzing other properties - the lexicon - of the text entered.

A quite similar code might produce the decreasing, or increasing, rearrangement of a function. Rearranging a function is just about the same as scrambling a piece of poetry in the mentioned way. Rearrangements act on real-valued functions of one or several real variables and result in functions whose level sets are endowed with special symmetries or special geometric properties. Typically, a rearranged function looks little appealingly. On the other hand, it shares key properties with, and simultaneously is simpler and more tractable than, the original function - a decisive gain, possibly.

Several types of rearrangements are known (a catalogue is in Kawohl [9]) and employed in various branches of analysis. Here we shall limit ourselves to present the so-called rearrangements à la Hardy & Littlewood.

§ 2. Definitions and basic properties

Let

G = a measurable subset of euclidean n-space $\mathbb{R}^n$,

u = a measurable real-valued function, defined in G

and suppose

m(G), the Lebesgue measure of G, is positive.

We are concerned with the following three objects:

name	symbol
distribution function of u	μ
decreasing rearrangement of u	u^*
symmetric rearrangement of u	$u^\bigstar$

Definition 1. μ is the map from $[0,\infty]$ into $[0,\infty]$ such that

(1) $$\mu(t) = m\{x \in G : |u(x)| > t\} .$$

In other words, $\mu(t)$, the value of μ at a point $t \geq 0$, is the measure of

$$(2) \qquad \{x \in G : |u(x)| > t\} ,$$

a level set of u.

The following properties hold:

1. μ is a decreasing function;
2. μ is right-continuous;
3. $\mu(t-) - \mu(t)$, the jump of μ at any $t \geq 0$, $= m\{x \in G : |u(x)| = t\}$;
4. $\mu(0)$ = measure of the support of u;
5. support of $\mu = [0, \text{ess sup } |u|]$.

In fig. 1 a typical situation is sketched.

Definition 2. u^* is the distribution function of μ. As μ decreases and is right-continuous, this definition can be rephrased in the following way:

$$(3a) \qquad \begin{aligned} u^*(s) &= \sup\{t \geq 0 : \mu(t) > s\} \\ &= \min\{t \geq 0 : \mu(t) \leq s\} \end{aligned}$$

for every $s \geq 0$.

The following properties hold:

1. u^* is a decreasing function;
2. u^* is right-continuous;
3. $u^*(0) = \text{ess sup } |u|$;
4. support of $u^* = [0, m(G)]$.

The following theorem, though easy, is basic: <u>The distribution function of</u> u^* <u>is exactly</u> μ; in other words, u <u>and</u> u^* <u>are equimeasurable or equidistributed</u>.

Corollaries:

1. The process of successively forming distribution functions stops at the second step.
2. Any operator, which acts in a space of measurable functions and whose effect depends on the content of level sets only, is <u>invariant</u> under rearrangements à la Hardy & Littlewood. For instance

$$\int_G A(|u(x)|)dx = \int_0^{m(G)} A(u^*(s))ds$$

whenever A is monotonic. In fact both integrals equal

$$\int_0^\infty A(t)\ [-d\mu(t)]\ .$$

Note incidentally that the last statement is a form of Cavalieri's principle.

3. The decreasing rearrangement u^* of u is characterized by the following pair:

First. u^* is a decreasing right-continuous map from $[0,\infty]$ into $[0,\infty]$.

Second. For every nonnegative t, the lenght of the level set $\{s \geq 0 : u^*(s) > t\}$ coincides with the n-dimensional measure of the level set $\{x \in G : |u(x)| > t\}$.

This characterization has a geometric meaning, which the author has tried to display in fig. 1.

Definition 3. $u^\bigstar$ is the nonnegative function, defined in $\mathbb{R}^n$ by

$$u^\bigstar(x) = u^*(C_n|x|^n)\ . \tag{4a}$$

Here $|x| = (x_1^2 + x_2^2 + \ldots + x_n^2)^{1/2}$ and C_n is the measure of the unit n-dimensional ball.

The following properties hold:

1. $u^\bigstar$ is a <u>radial</u> function, i.e. invariant under rotations about the origin;
2. u and $u^\bigstar$ are equidistributed;
3. The level set

$$\{x \in \mathbb{R}^n : u^\bigstar(x) > t\}$$

is the ball whose center is the origin and whose measure equals the measure of

$$\{x \in G : |u(x)| > t\}\ ;$$

4. The support of $u^\bigstar$ is G, the ball whose center is the origin and whose measure equals the measure of G.

§ 3. Alternative formulas.

The preceding matter can be recast in a more compact form. Recall that any nonnegative integrable function f is the superimposition of the characteristic functions of its level sets, i.e.

$$(5) \qquad f = \int_0^\infty \mathbb{1}_{\{x\in\mathbb{R}^n : f(x)>t\}} dt \quad ,$$

where $\mathbb{1}$ stands for characteristic function and the integral is Bochner's. Suppose u is integrable over G. Then a collage gives:

$$(3b) \qquad u^* = \int_0^\infty \mathbb{1}_{[0,\mu(t)]} dt$$

$$(4b) \qquad u^\bigstar = \int_0 \mathbb{1}_{\{x\in\mathbb{R}^n : C_n |x|^n < \mu(t)\}} dt \quad .$$

These formulas have various easy consequences. For instance, (3b) tells us that

$$- \frac{du^*}{ds} \quad ,$$

the derivative of $-u^*$ in the sense of distributions, is the following linear form:

$$C_0^\infty(]0,\infty[) \ni \varphi \longrightarrow \int_0^\infty \varphi(\mu(t))dt \quad .$$

§ 4. Key theorems.

In most applications the main theorems on rearrangements à la Hardy & Littlewood are theorems A, B, C below.

Theorem A. Suppose G is a measurable subset of $\mathbb{R}^n$, u and v are nonnegative measurable functions. Then

$$(10) \qquad \int_G u(x)v(x)dx \leq \int_0^{m(G)} u^*(s)v^*(s)ds \quad .$$

Theorem B. Suppose f, g, h are nonnegative measurable functions. Then

$$(11) \qquad \int_{\mathbb{R}^n} dx \int_{\mathbb{R}^n} f(x)g(y)h(x-y)dy \le \int_{\mathbb{R}^n} dx \int_{\mathbb{R}^n} f^\star(x)g^\star(y)h^\star(x-y)dy \quad .$$

Theorem C. Suppose G is an open subset of $\mathbb{R}^n$, u belongs to Sobolev space $W_0^{1,p}(G)$, $p \ge 1$. Then

$$(12) \qquad \int_G |\text{grad } u|^p dx \ge \int_{G^\star} |\text{grad } u^\star|^p dx \ .$$

Theorem A is by Hardy & Littlewood. Proofs are e.g. in [7]. Theorem A is simple, but crucial. One may claim that most theorems from real and harmonic analysis, which can be demonstrated via rearrangements, are consequences of theorem A.

Theorem B is due to F. Riesz. Proofs are in Riesz's paper [14], and in [7]. Generalizations and improvements are in [2],[12].

Theorem C tells us that total variation - and related things - improve with symmetric rearrangements. Theorem C might be called Polya & Szegö principle. Polya, Szegö, and their followers, derived a lot of isoperimetric theorems and isoperimetric properties of eigenvalues (e.g. Faber & Krahn theorem on the principal frequency of a membrane, Poincaré inequality for capacity, St. Venant principle for torsional rigidity) from theorem C. Cf. [13]. By the way, theorem C is crucial for settling Sobolev inequality in a sharp form, cf. [1],[19].

The proof of theorem C is based on ideas, that Polya & Szegö mastered and ultimately are to be found in papers [4],[10] by Faber and Krahn. However, implementing these ideas needs tools from geometric measure theory, that have been set up in recent times only. Exhaustive proofs of theorem C are offered in [8],[16],[17],[18],[19]. An ingenious proof of the $p = 2$ case is due to Lieb [11]. It should be stressed that, after all, theorem C is a consequence of the isoperimetric theorem in $\mathbb{R}^n$.

A last (but not least) remark. Functions exist, which really differ from their symmetric rearrangement and render (12) an equality. Kawohl [9] and Friedman & McLeod [5] showed that equality in (12) implies $u = u^\bigstar$ if the set of critical points of u is (in a suitable sense) thin enough.

§ 5. Algorithms.

How to compute a rearrangement of a function? Only in few bookish examples one can calculate by hands and conclude with a closed formula. In general an algorithm is needed.

In this section we present an algorithm for computing (and plotting) the decreasing rearrangement of function of one real variable. The algorithm has its roots in a 1899 paper [15] by C. Somigliana and is based on the observation that the decreasing rearrangement of a step (i.e. piecewise constant) function can be rigorously computed via combinatoric arguments only. More precisely, let us emphasize the following lucky fact: the decreasing rearrangement of a step function of one real variable is the step function which is obtained merely by assembling the original blocks in order of decreasing height. See fig. 3.

Let u be a real-valued function defined in an interval $[a,b]$. Suppose u is smooth enough; for instance, u is absolutely continuous and its derivative u' is in $L^q(a,b)$ for some $q \geq 1$.

Consider the step function, that we name stpfn, defined as follows:

$$\text{stpfn} = \sum_{i=1}^{n} u\left(\frac{x_{i-1}+x_i}{2}\right) \mathbb{1}_{[x_{i-1}, x_i[} \quad .$$

Here $x_0, x_1, \ldots, x_n$ is a partition of $[a,b]$ made by equidistant point, i.e.

$$x_i = a + i(b-a)/n \qquad (i = 0,1,\ldots,n) \; .$$

The following inequality holds:

$$\|u - \mathrm{stpfn}\|_{L^p(a,b)} \le C(h/2)^{1+1/p-1/q} \|u'\|_{L^q(a,b)} ,$$

where p is any exponent $\ge q$,

$$C = \left(1 + \frac{q}{p(q-1)}\right)^{1/q} \left(1 + p\left(1 - \frac{1}{q}\right)\right)^{-1/p} \frac{p}{B(1/p, 1-1/q)} ,$$

B stands for Euler's beta function, and

$$h = (b-a)/n .$$

Thus stpfn is arbitrarily close to u in the $L^q(a,b)$-metric, if n is large enough.

On the other hand, a theorem by Chiti [3] tell us that the decreasing rearrangement is a non-expansive map from $L^p(a,b)$ into $L^p(0,l)$, where $l = b-a$. Then

$$\|u^* - (\mathrm{stpfn})^*\|_{L^p(0,l)} \le \|u - \mathrm{stpfn}\|_{L^p(a,b)} .$$

In conclusion, an algorithm for computing u^* is as follows:

(i) (stpfn)*, something we can compute easily, is an approximation of u*, the thing we want to compute;

(ii) the quality of such an approximation is decided by the mesh size h, according to the following estimate:

$$\|u^* - (\mathrm{stpfn})^*\|_{L^p(0,l)} \le C(h/2)^{1+1/p-1/q} \|u'\|_{L^q(a,b)} .$$

The algorithm can be implemented by the code below. Fig. 4 may help to grasp the situation. Various examples, which have been worked out via our algorithm and code, are shown in figs. 5 to 8.

Code for rearranging a function (<u>one</u> real variable, closed form) à la Hardy & Littlewood.

Language: Expanded CBM V2 BASIC.

	10-20: Input of data.
10 DEF FNU(X) =	Enter function, name the dummy variable x . The function is named fnu .
15 A= : B=	Enter end points.
20 N=	Enter the number of points, where fnu has to be sampled
	50-100: Function fnu is sampled and the sample values of fnu are arranged in decreasing order. The largest sample value and the smallest sample value are singled out.
50 DIM V(N) : V(0)=1E+30	Inform that v is a list of numbers, having (n+1) entries. The first entry of v is defined to be $+\infty$.
55 FOR I=1 TO N	For $i=1,2,\dots,n$ suppose the entries of v no. 0 to (i-1) are specified and v(0),....,v(i-1) is an arrangement in decreasing order. Do 60-85.
	Segment 60-85 specifies v(i), the i-th entry of v, and up-dates the entries of v no. 0 to (i-1). Segment 60-85 results in the decreasing rearrangement of earlier values of v(0),...,v(i-1) and the value of fnu at the mid of point of $[x_{i-1},x_i]$. Here $x_i = a+i(b-a)/n$.
60 X=A+(I-.5)*(B-A)/N : Y=FNU(X)	Pick the mid point of the interval $[x_{i-1},x_i]$. Evaluate fnu at this point and save the result under the name y .

`65 J=I-1`	Start scanning the list v(0),...,v(i-1) backwards.
`70 IF V(J) > =Y THEN 85`	Test: does the entry at hand exceed y ?
`75 V(J+1) = V(J)` `80 J=J-1 : GOTO 70`	Relabel entries, and go on scanning, as long as the test is negative, i.e. until an entry is found which exceeds y .
`85 V(J+1)=Y`	Enroll y immediately after the earliest entry whith exceeds y . Stop scanning as soon as the test is positive, i.e. an entry is met, which exceeds y .
`90 NEXT I`	
`100 MAX=V(1) : MIN=V(N)`	
`30 PRINT CHR$(147)`	
`90 PRINT I;TAB(10)X;TAB(25)X`	Display sample points and sample values.
`105 PRINT CHR$(147)`	
`110 FOR I=1 TO N : PRINT I,V(I) : NEXT`	Display the sample values in decreasing order.

200-300: Plotting stpfn and (stpfn)* .

Here $$\text{stpfn} = \sum_{i=1}^{n} \text{fnu}\left(\frac{x_{i-1}+x_i}{2}\right) \mathbb{1}_{[x_{i-1},x_i[} .$$

```
200 PRINT CHR$(147)
205 INPUT "height";H : INPUT "lenght";L
206 REM Enter scale factors. H must satisfy  0 < H ≤ 199.  L should exceed  N .

210 HIRES 0,1 : POKE 53280,1
215 LINE 0,100-H/2,0,100+H/2,1
```

```
220  J=0 : X=A

225  I=1+INT((X-A)/(B-A)*N)
230  Y=FNU(A+(I-.5)*(B-A)/N)  : Y=100+H*((MAX+MIN)/2-Y)/(MAX-MIN)
235  PLOT J,Y,1
240  Y=V(I) : Y=100+H*((MAX+MIN)/2-Y)/(MAX-MIN)
245  PLOT J,Y,1
250  PLOT J,100+H/2,1

255  J=J+1 : IF  J < 320  THEN 270
260  GET C$ : IF C$="" THEN 260
265  J=0 : HIRES 0,1
270  X=X+(B-A)/L : IF X < B THEN 225

275  LINE J,100+H/2,J,100-H/2,1
290  GET C$ : IF C$="" THEN 290


300  END
```

§ 5. Applications of rearrangements, a very short account

Rearrangements of functions are fairly often employed in real and functional analysis, in theorems on singular integrals, in some topics from the calculus of variations.

Rearrangements intervene also in second-order (linear and nonlinear) elliptic partial differential equations. Consider a second-order elliptic partial differential equation, whose leading part has a divergence structure. Experience has shown that the very architecture of such an equation virtually comprises crucial information about level sets of solutions. Thus realistic and explicit estimates of solutions can be cogently derived by decoding this information. Rearrangements are an ad hoc key for such a decoding.

Papers [20] to [61] trend, more or less, this way. Below we sketch a simple example, patterned on [55],[60].

Consider the following problem:

$$\text{(20a)} \qquad \begin{cases} - \sum\limits_{i,j=1}^{n} \dfrac{\partial}{\partial x_i} \{a_{ij}(x) \dfrac{\partial y}{\partial x_j}\} + c(x)u = f(x) \quad \text{in } G \\ \\ u = 0 \quad \text{on} \quad \partial G \ , \end{cases}$$

where G is an open subset of $\mathbb{R}^n$, the measure $m(G)$ of G is finite and the following ellipticity conditions hold:

$$\text{(20b)} \qquad \sum_{i,j=1}^{n} a_{ij}(x)\xi_i\xi_j \geq \sum_{i=1}^{n} \xi_i^2 \quad \text{and} \quad c(x) \geq 0 \ .$$

Suppose f belongs to $L^p(G)$, $p = 2n/(n+2)$, and let u be the solution to (20a) from Sobolev space $W_0^{1,2}(G)$ (our hypothesis on the right-hand side is just the appropriate one for a solution to exist in the specified function space).

The following statements hold.

First. The distribution function μ of solution u satisfies:

$$\text{(21)} \qquad 1 \leq [-\mu'(t)]\ \kappa_n^2 \mu(t)^{2/n-2} \int_0^{\mu(t)} f^*(s)ds$$

for almost every $t > 0$. Here $\kappa_n = n^{-1} C_n^{-1/n}$, the isoperimetric constant of $\mathbb{R}^n$.

Second. The following inequalities hold:

$$(22) \quad \begin{cases} u^{\bigstar} \le v \\ \int_G |\text{grad } u|^q dx \le \int_{G^{\bigstar}} |\text{grad } v|^q dx \quad \text{for} \quad 0 < q \le 2 , \end{cases}$$

where $G^{\bigstar}$ is the ball having the same measure as G and

$$(23) \qquad v(x) = n^{-2} C_n^{-2/n} \int_{C_n |x|^n}^{m(G)} dr \, r^{-2+2/n} \int_0^r f^*(s) ds \quad .$$

The isoperimetric theorem in $\mathbb{R}^n$ is an essential ingredient for the proof of (21). We stress that no hypothesis on the data is involved in the proof of (21), except ellipticity conditions (20b) and the stated hypothesis on the right-hand side.

The second statement is a consequence of the first. Note that v is the $W_0^{1,2}(G^{\bigstar})$-solution to

$$(24) \quad \begin{cases} -\Delta v = f & \text{in } G^{\bigstar} \\ v = 0 & \text{on } \partial G^{\bigstar} , \end{cases}$$

the simplest problem having type (20). Conclusions:

$$\|u\| \qquad \text{and} \qquad \int_G |\text{grad } u|^q dx$$

are a maximum when the relevant data - the differential operator, the ground domain, the righ-hand side - are as simple as possible - the Laplace operator, a ball, a radial function. Here $\| \;\; \|$ stand for the norm in any Luxemburg-Zaanen space (so Banach space of measurable real-valued functions are called, whose norm is invariant under rearrangements à la Hardy & Littlewood; note that Lebesgue, Lorentz, Orlicz spaces fall into this category).

Thus estimating a Luxemburg-Zaanen norm, or a Dirichlet-type integral, of the solution to a problem having type (20) amounts to the considerably simpler and well-defined task of estimating the solution of problem (24) - a solution which is explicitly represented by formula (23).

Below we list some estimates of the solution to problem (20), which can be derived via the preceding arguments. Let us stress that all these estimates are sharp.

(i) $\max|u| \le C\,[m(G)]^{2/n-1/p}\,\|f\|_{L^p(G)}$,

where $p > n/2$,

$$C = \frac{\Gamma(n/2+1)^{2/n}}{n(n-2)\pi}\left[\frac{\Gamma(1+p')\,\Gamma(n/(n-2)-p')}{\Gamma(n/(n-2))}\right]^{\frac{1}{p'}}$$

and $p' = p/(p-1)$. This estimate was proved by H. Weinberger in "Symmetrization in uniformly elliptic problems", Studies in Math. Anal., Stanford Univ. Press, 1962.

(ii) $$\|u\|_{L(q,k)} \le \frac{q^2\,\Gamma(1+n/2)^{2/n}}{(q-1)n^2\pi}\,\|f\|_{L(p,k)} ,$$

where $1 < p < n/2$, $k \ge 1$, $q = np/(n-2p)$ and $L(p,k)$ denotes a Lorentz space.

(iii) $$\left(\int_G |\mathrm{grad}\, u|^q dx\right)^{1/q} \le C\left(\int_G |f|^p dx\right)^{1/p} ,$$

where $1 < p < 2n/(n-2)$, $q = np/(n-p)$ and

$$C = \frac{q^{-1/q}}{n\sqrt{\pi}}\left(\frac{p}{p-1}\right)^{1/p}\left[\frac{\Gamma(n)\,\Gamma(n/2)}{2\Gamma(n/p)\,\Gamma(n-n/p)}\right]^{1/n} .$$

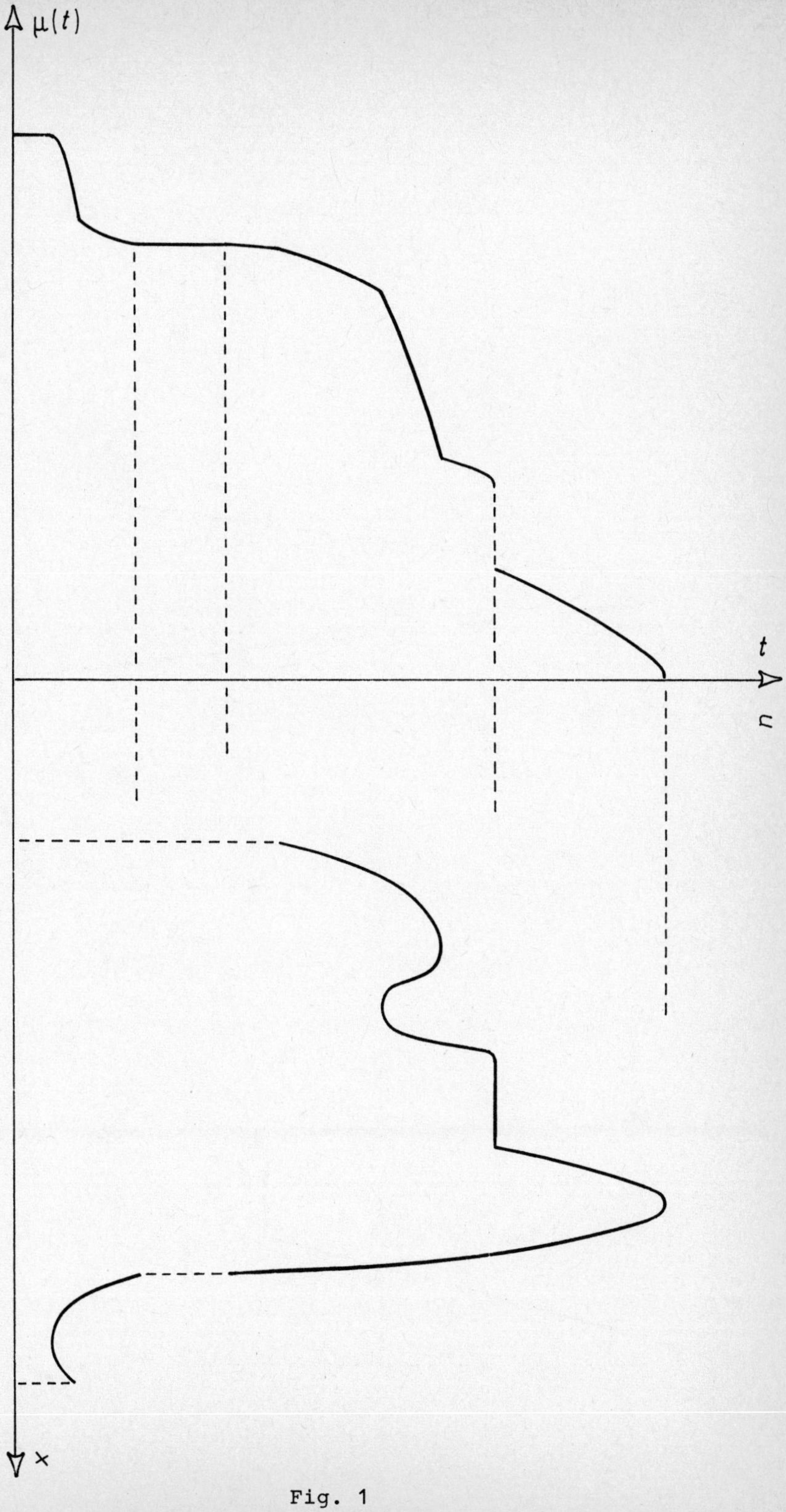

Fig. 1

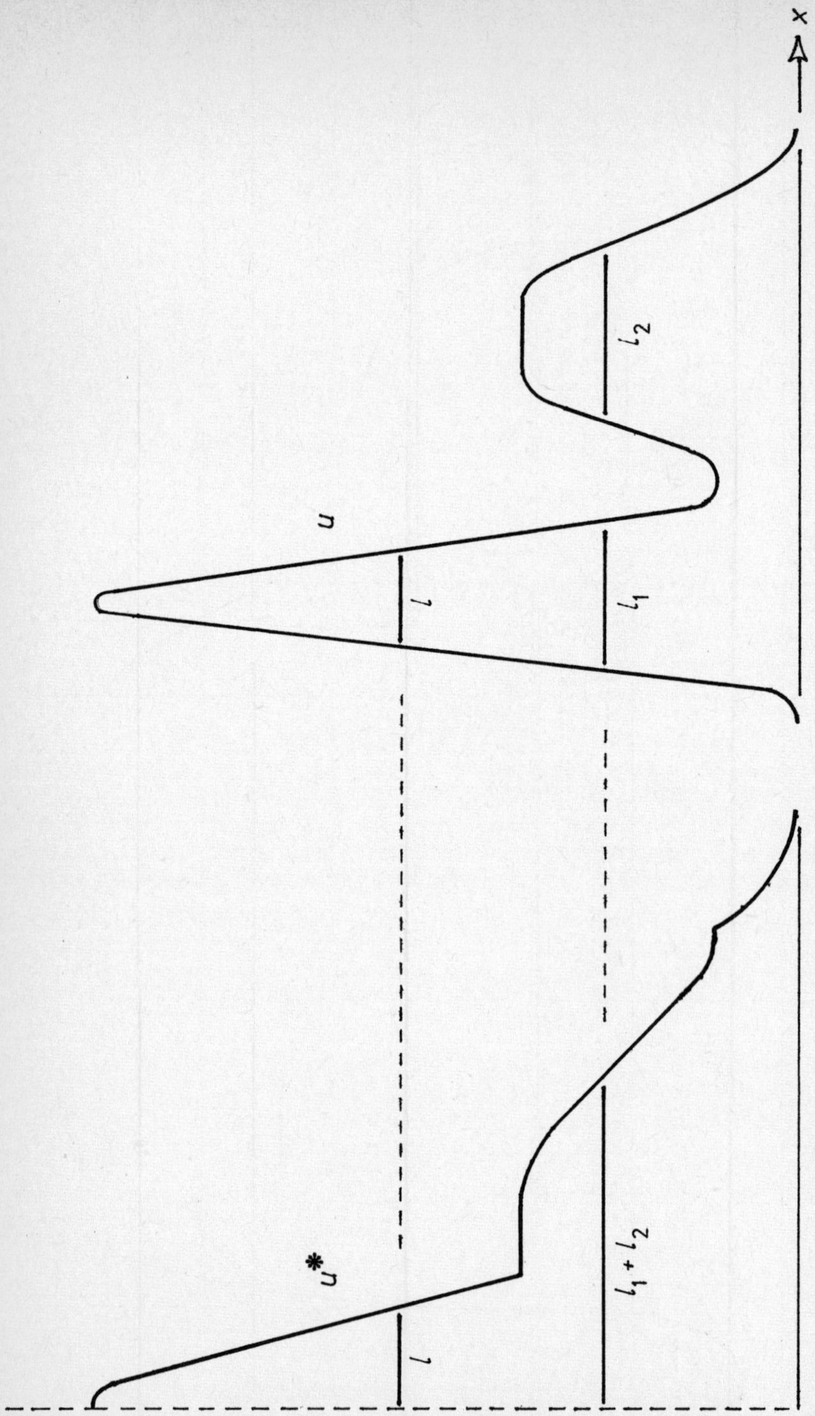

Fig. 2

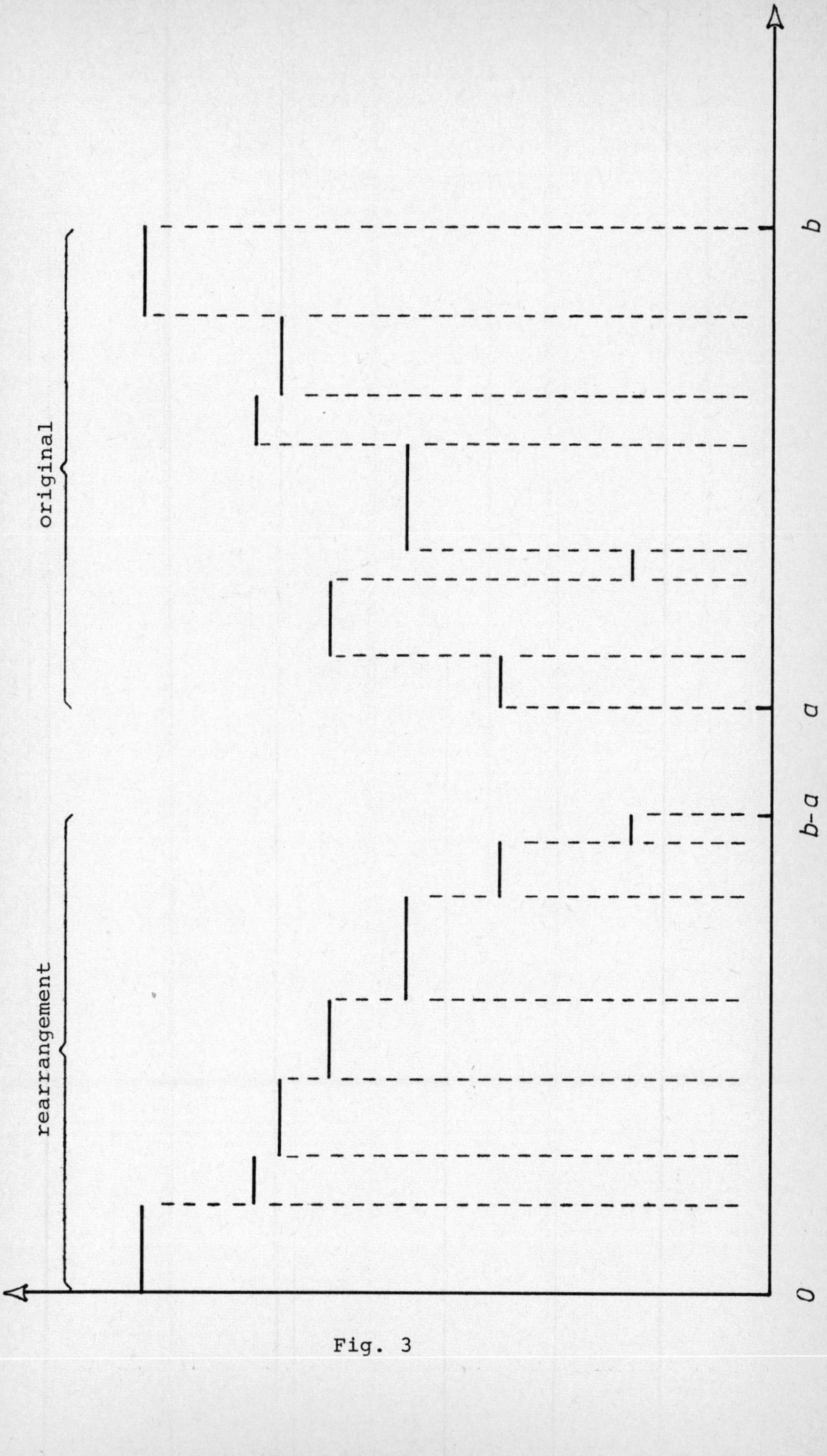

Fig. 3

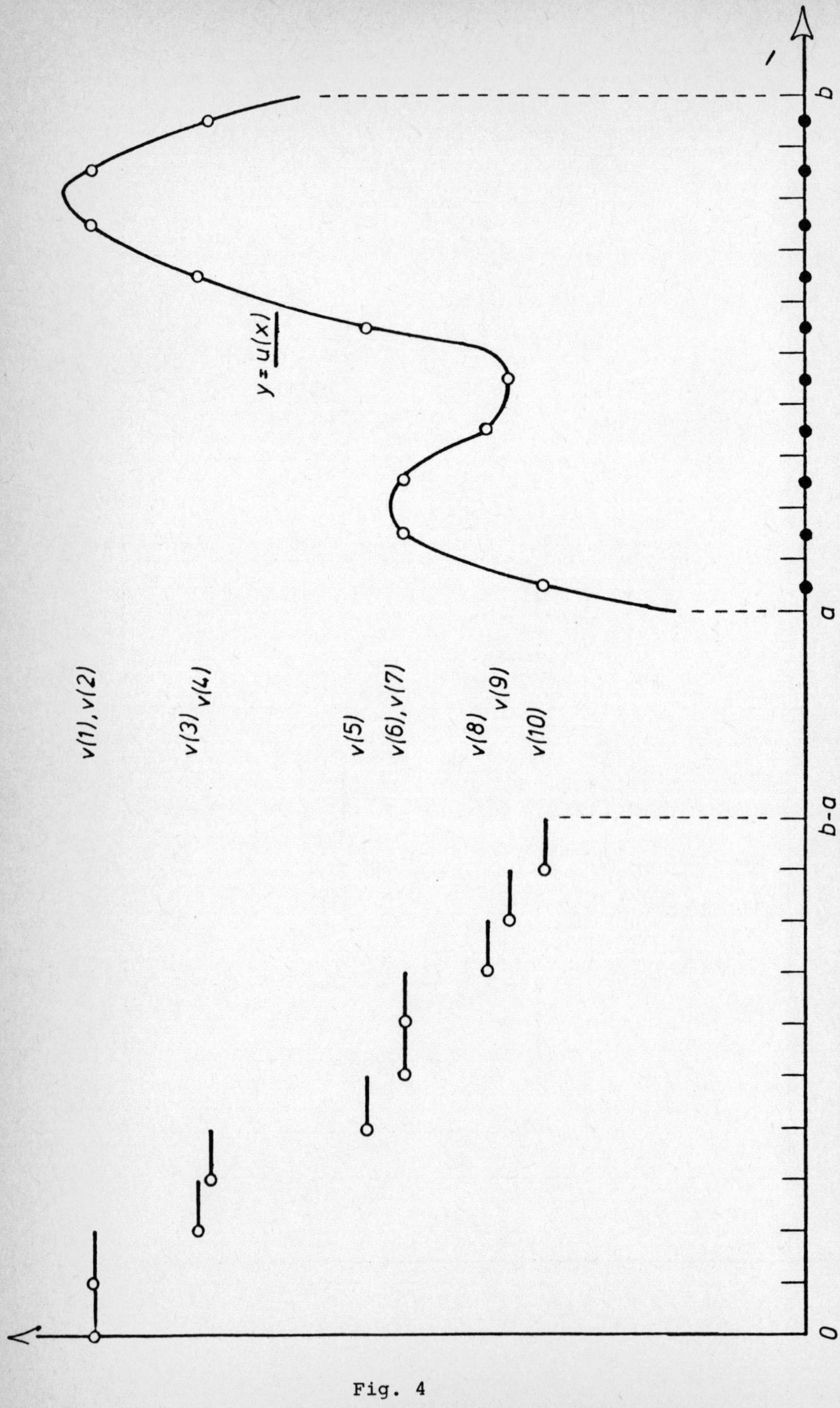
y = u(x)
a
b
x0
x1
x2
x3
x4
x5
x6
x7
x8
x9
x10
0
b-a
s0
s1
s2
s3
s4
s5
s6
s7
s8
s9
s10
v(1), v(2)
v(3) v(4)
v(5)
v(6), v(7)
v(8) v(9)
v(10)

Fig. 4

$$u(x) = (4x^4 - 5x^2 + 1)^2 \quad , \qquad -1 \le x \le 1$$

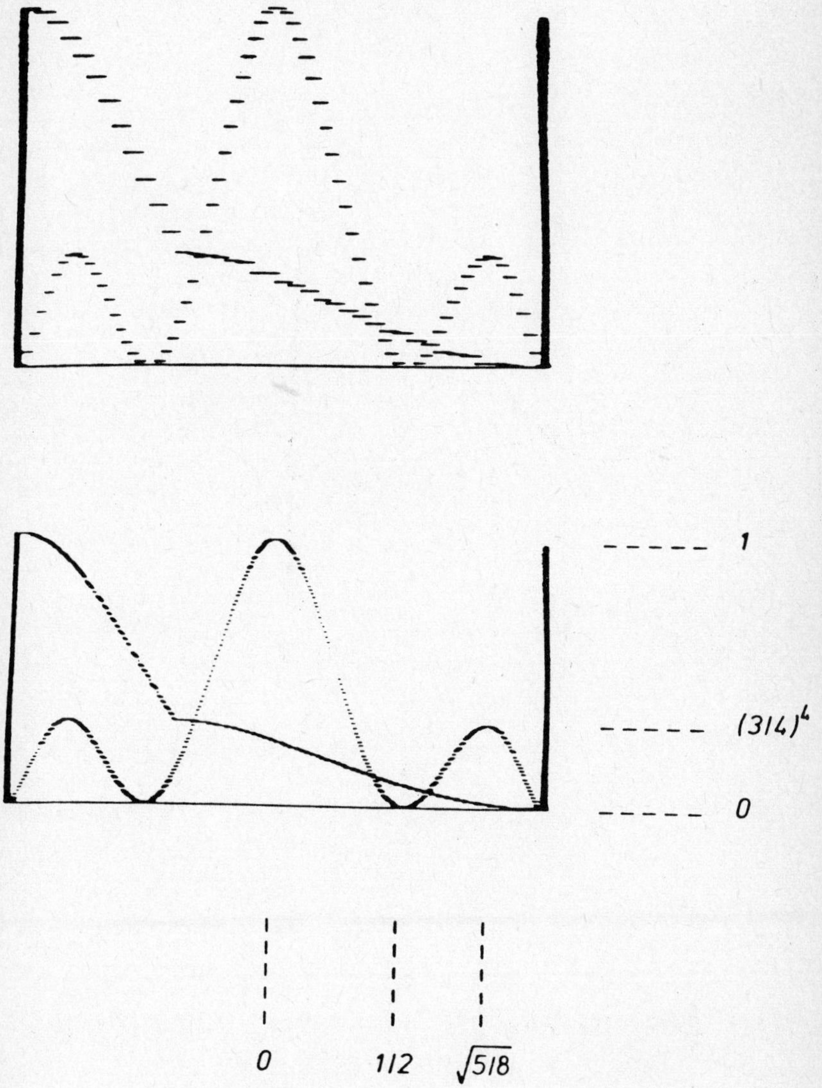

Fig. 5

$$u(x) = 468 + 185x^2 + 65x^3 - 45x^4 - x^5 \quad , \quad -2 \leq x \leq 3$$

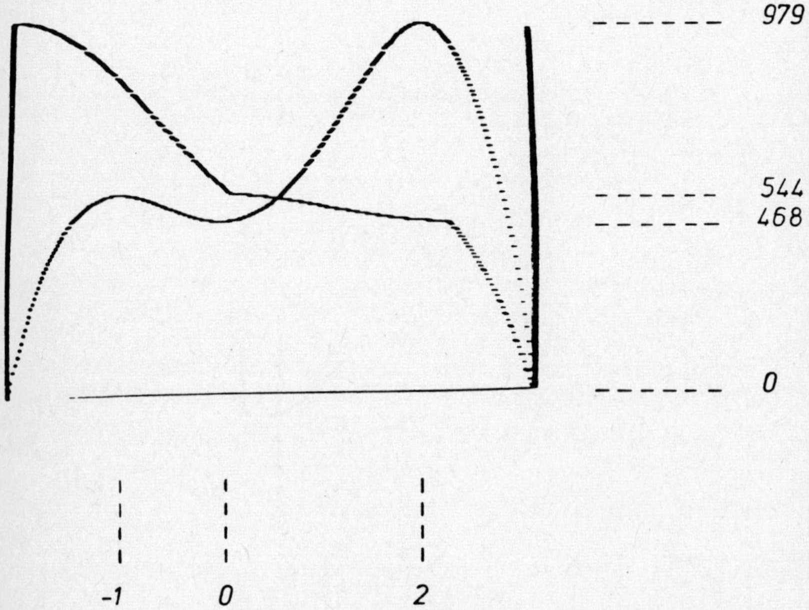

$$u'(x) = -5x(x+1)(x-2)(x+37)$$

Fig. 6

$$u(x) = 3\sin x + \sin(4x) \qquad , \qquad 0 \le x \le \pi$$

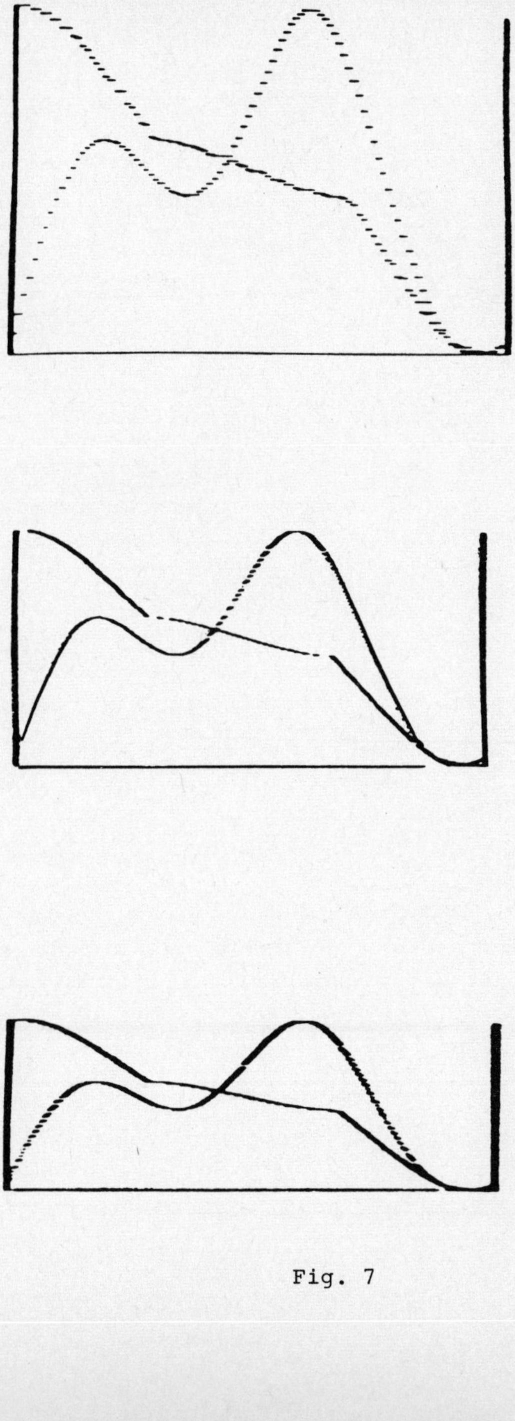

Fig. 7

$$u(x) = 8\sin x - 4\sin(3x) + 2\sin(9x) - \sin(27x) \quad , \quad 0 \le x \le \pi$$

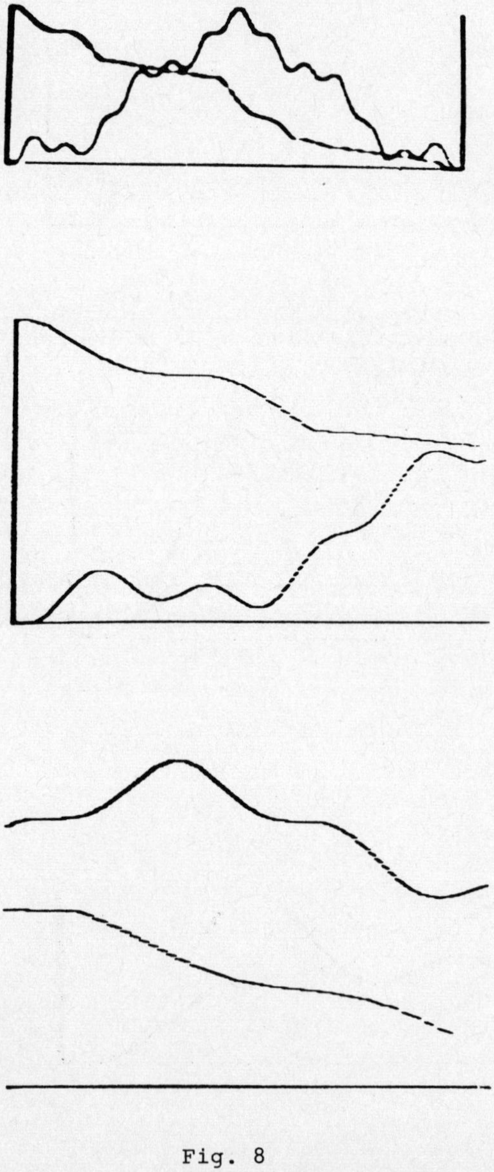

Fig. 8

REFERENCES

[1] T. Aubin, Problèmes isopérimetriques et espaces de Sobolev (J. Diff. Geometry, 11, 1976).

[2] H.J. Brascamp - E.H. Lieb - J.M. Luttinger, A general rearrangement inequality for multiple integrals (J. Functional Anal. 17, 1974).

[3] G. Chiti, Rearrangements of functions and convergence in Orlicz spaces (Appl. Anal. 9, 1979).

[4] G. Faber, Beweis dass unter allen homogenen membranen von gleicher fläche und gleicher spannung die kreisförmige den tiefsten grundton gibt (Sitzungsber. Bayer. Akad. Wiss., Math.-Naturwiss.Kl., 1923).

[5] A. Friedman - B.McLeod, Strict inequalities for integrals of decreasingly rearranged functions (to appear).

[6] A.M. Garsia - E. Rodemich, Monotonicity of certain functionals under rearrangements (Ann. Inst. Fourier Grenoble 24, 1974).

[7] Hardy-Littlewood-Polya, Inequalities (Cambridge Univ. Press, 1964).

[8] K. Hildén, Symmetrization of functions in Sobolev spaces and the isoperimetric inequality (Manuscripta Math. 18, 1976).

[9] B. Kawohl, Rearrangements and convexity of level sets in PDE (Lecture Notes in Math. 1150, Springer Verlag 1985).

[10] E. Krahn, Über eine von Rayleigh formulierte Minimaleigenschaft des Kreises (Math. Ann. 94, 1924).

[11] E.H. Lieb, Existence and uniqueness of the minimizing solution of Choquard's nonlinear equation (Studies in Appl. Math. 57, 1977).

[12] E.H. Lieb, Sharp constants in the Hardy-Littlewood-Sobolev and related inequalities (Ann. of Math. 118, 1983).

[13] G. Polya-G.Szegö, Isoperimetric inequalities in mathematical physics (Princeton Univ. Press, 1951).

[14] F. Riesz, Sur une inegalité intégrale (J. London Math. Soc. 5, 1930).

[15] C. Somigliana, Sulle funzioni reali d'una variabile, Considerazioni sulle funzioni ordinate (Rendiconti R. Accademia dei Lincei, vol.8, 1899).

[16] E. Sperner, Zur Symmetrisierung für Funktionen auf Sphären (Math. Z. 134, 1973).

[17] E. Sperner, Symmetrisierung für Funktionen mehrerer reeller Variablen (Manuscripta Math. 11, 1974).

[18] W. Spiegel, Über die Symmetrisierung stetiger Funktionen im euklidischen Raum (Archiv der Math. 24, 1973).

[19] G. Talenti, Best constant in Sobolev inequality (Ann. Mat. Pura Appl. 110, 1976).

[20] A. Alvino, Formule di maggiorazione e regolarizzazione per soluzioni di equazioni ellittiche del secondo ordine in un caso limite Rend. Acc. Naz. Lincei, ser.8, vol. 62, 1977).

[21] A. Alvino - G. Trombetti, Equazioni ellittiche con termini di ordine inferiore e riordinamenti (Rend. Acc. Naz. Lincei, ser.8, vol. 66, 1979).

[22] A. Alvino - G. Trombetti, Sulle migliori costanti di maggiorazione per una classe di equazioni ellittiche degeneri (Ricerche di Mat. 27, 1979).

[23] A. Alvino - G. Trombetti, Su una classe di equazioni ellittiche non lineari degeneri (Ricerche di Mat. 29, 1980).

[24] A. Alvino - G. Trombetti, Sulle migliori costanti di maggiorazione per una classe di equazioni ellittiche degeneri e non (Ricerche di Mat. 30, 1981).

[25] A. Alvino - G. Trombetti, A lower bound for the first eigenvalue of an elliptic operator (J. Math. Analysis and Applications 94, 1983).

[26] A. Alvino - P.L. Lions - G. Trombetti, A remark on comparison results via symmetrization (in corso di stampa).

[27] C. Bandle, Bounds for the solutions of boundary value problems (J. Math. Anal. Appl. 54, 1976).

[28] C. Bandle, On symmetrizations in parabolic equations (J. Analyse Math. 30, 1976).

[29] C. Bandle, Estimates for the Green's functions of elliptic operators (SIAM J. Math. Anal. 9, 1978).

[30] C. Bandle - J. Mossino, Application du réarrangement à une inéquation variationnelle (C.R.A.S. Paris, t.296, Série I, 1983).

[31] C. Bandle - R.P. Sperb - I. Stakgold, The single, steady-state irreversible reaction (in corso di stampa).

[32] P. Buonocore, Sulla simmetrizzazione in equazioni paraboliche degeneri (Boll. U.M.I., (6) 3-B, 1984).

[33] G. Chiti, Norme di Orlicz delle soluzioni di una classe di equazioni ellittiche (Boll. U.M.I. (5) 16-A, 1979).

[34] G. Chiti, An isoperimetric inequality for the eigenfunctions of linear second order elliptic operators (Boll. U.M.I. (6) A-1,1982).

[35] G. Chiti, A reverse Hölder inequality for the eigenfunctions of linear second order elliptic equations (Z.A.M.P. 33, 1982).

[36] P.S. Crooke - R.P. Sperb, Isoperimetric inequalities in a class of nonlinear eigenvalue problems (SIAM J. Math. Anal. 9, 1978).

[37] E. Giarrusso - D. Nunziante, Symmetrization in a class of first-order Hamilton-Jacobi equations (Nonlinear Analysis T.M.A. 8,1984).

[38] E. Giarrusso - D. Nunziante, Comparison theorems for a class of first-order Hamilton-Jacobi equations (in corso di stampa).

[39] B. Kawohl, On the isoperimetric nature of a rearrangement inequality and its consequences for some variational problems (LCDS Report 84-4, Providence 1984).

[40] P. Laurence - E.W. Stredulinsky, A new approach to queer differential equations (to appear in Comm. Pure Appl. Math.)

[41] P.L. Lions, Quelques remarques sur la symétrisation de Schwarz (Nonlinear partial differential equations and their applications, Collège de France Seminars, vol. 1, Pitman 1981).

[42] C. Maderna, Optimal problems for a certain class of nonlinear Dirichlet problems (Boll. U.M.I., Suppl.1, 1980).

[43] C. Maderna, On level sets of Poisson integrals in disks and sectors (Boll. U.M.I. (6) ser. C-2, 1983).

[44] C. Maderna - S. Salsa, Symmetrization in Neumann problems (Appl. Anal. 9, 1979).

[45] C. Maderna - S. Salsa, A priori bounds in nonlinear Neumann problems (Boll. U.M.I. (5) 16-B, 1979).

[46] C. Maderna - S. Salsa, Sharp estimates for solutions to a certain type of singular elliptic boundary value problem in two dimensions (Appl. Anal. 12, 1981).

[47] C. Maderna - S. Salsa, Some special properties of solutions to obstacle problems (Rend. Sem. Mat. Univ. Padova 71, 1984).

[48] J. Mossino, A priori estimates for a model of Grad-Mercier type in plasma confinement (Appl. Anal. 13, 1982).

[49] J. Mossino, A generalization of the Payne-Rayner isoperimetric inequality (Boll. U.M.I. (6) A-2, 1983).

[50] J. Mossino, Inégalités isopérimetriques et applications en physique (Herman, 1984).

[51] J. Mossino - J.M. Rakotoson, Isoperimetric inequalities in parabolic equations (to appear).

[52] J. Mossino - R. Temam, Directional derivative of the increasing rearrangement mapping and application to a queer differential equation in plasma physics (Duke Math. J. 48, 1981).

[53] R.P. Sperb, Maximum principles and their applications (Academic Press, 1981).

[54] E. Sperner, Spherical symmetrization and eigenvalue estimates (Math. Z. 176, 1981).

[55] G. Talenti, Elliptic equations and rearrangements (Ann. Scuola Norm. Sup. Pisa (4) 3, 1976).

[56] G. Talenti, Nonlinear elliptic equations, rearrangements of functions and Orlicz spaces (Ann. Mat. Pura Appl. (4) 120, 1979).

[57] G. Talenti, Some estimates for solutions to Monge-Ampère equations in dimension two (Ann. Scuola Norm. Sup. Pisa (4) 8, 1981).

[58] G. Talenti, On the first eigenvalue of the clamped plate (Ann. Mat. Pura Appl. (4) 129, 1981).

[59] G. Talenti, A note on the Gauss curvature of harmonic and minimal surfaces (Pacific J. Math. 101, 1982).

[60] G. Talenti, Linear elliptic p.d.e.'s: level sets, rearrangements and a priori estimates of solutions (to appear).

[61] J.L. Vazquez, Symétrisation pour $u_t = \Delta\varphi(u)$ et applications (C.R.A.S.P., t.295, 1982).

FONDAZIONE C.I.M.E.
CENTRO INTERNAZIONALE MATEMATICO ESTIVO
INTERNATIONAL MATHEMATICAL SUMMER CENTER

"Inverse Problems"

is the subject of the First 1986 C.I.M.E. Session.

The Session, sponsored by the Consiglio Nazionale delle Ricerche and the Ministero della Pubblica Istruzione, will take place under the scientific direction of Prof. GIORGIO TALENTI (Università di Firenze) at Villa «La Querceta», Montecatini Terme (Pistoia), Italy, ***from May 28 to June 5, 1986.***

Courses

a) ***Inverse Eigenvalue Problems.*** (6 lectures in English).
Prof. Victor BARCILON (University of Chicago).

— Introduction. Eigenvalue problems arising in the study of vibrating systems: string, beam, elastic earth. Impulse response; equivalent data sets. Asymptotic behaviour of eigensolutions.
— Oscillatory kernels and Chebyshev systems. The work of Kellogg, M.G. Krein, Karlin.
— Inverse Sturm-Liouville problem. Borg uniqueness theorem. Improvements of Levinson and Marchenko. End-point formules.
— Inverse problem for the vibrating string. Discretization and construction algorithms. Existence results. End-point formulas and reduction to quadratures.
— Inverse problem for the vibrating beam. Discretization: oscillatory matrices.
— Inverse problem for the vibrating beam. Existence theorem.

References

- BARCILON, V., Inverse problem for the vibrating beam in the free-clamped configuration, Phil. Trans. Roy Soc. London, A 304 (1982), 211-251.
- BARCILON, V., Explicit solution of the inverse problem for a vibrating string, J. Math. Anal. Appl., 93 (1983), 222-234.
- BORG, G., Eine Umkehrung der Sturm-Liouvilleschen Eigenwertaufgabe, Acta Math., 78 (1946), 1-96.
- GANTMAKHER, F.P., The Theory of Matrices. Chelsea, New York, 1959.
- GANTMAKHER, F.P., and KREIN, M.G., Oscillation Matrices and Kernels and Small Vibrations of Mechanical Systems. Office of Technical Services, Washington, D.C., 1961.
- GELFAND, I.M., and LEVITAN, B.M., On a simple identity for the eigenvalues of a differential operator of the second order, Dokl. Akad. Nauk SSSR, 88 (1953), 593-596.
- GLADWELL, G.M. L., The inverse problem for the vibrating beam, Proc. Roy. Soc. London, A 393 (1984), 277-295.
- KARLIN, S., Total positivity. Stanford Univ. Press, 1968.
- KELLOGG, O.D., The oscillations of functions of an orthogonal set, Amer. Math. Soc., 38 (1916), 1-5.
- KELLOGG, O.D., Orthogonal function sets arising from integral equations, Amer. Math. Soc., 40 (1918), 145-154.
- KREIN, M.G., Sur les vibrations propres des tiges dont l'une des extremités est encastrée et l'autre libre, Comm. Soc. Math. Kharkoff, 12 (1935), 3-11.
- KREIN, M.G., On inverse problems for an inhomogeneous string, Dokl. Akad. Nauk SSSR, 82 (1952), 669-672 (in Russian).
- KREIN, M.G., Some new problems in the theory of Sturmian systems, Prik. Mat. Mekh., 16 (1952), 555-568 (in Russian).
- LEVINSON, N., The inverse Sturm-Liouville problem, Mat. Tidsskr. B (1949), 25-30.
- MARCHENKO, V.A., Some questions in the theory of one-dimensional linear differential operators of the second order, Amer. Math. Soc. Trans., 101 (1973), 1- 104.

b) ***Regularization Methods for Linear Inverse Problems.*** (6 lectures in English).
Prof. Mario BERTERO (Università di Genova).

Contents: Linear inverse problems and linear operator equations. Examples of ill- posed inverse problems. Generalized inverses of linear operators. Approximate solutions satisfying prescribed constraints. Ivanov constrained pseudosolutions and Miller least squares method. Regularizing algorithms: general definition and examples (Tikhonov regularizer, spectral windows etc.). Choice of the regularization parameter and the discrepancy principle. Stability estimates. Linear inverse problems with discrete data. Applications to first kind Fredholm integral equations.

Basic literature:

- GROETSCH, C.W., Genralized Inverses of Linear Operators. Dekker, New York, 1977.
- GROETSCH, C.W., The Theory of Tikhonov Regularization for Fredholm Equations of the First kind. Pitman, Boston, 1984.
- LAVRENTIEV, M.M., Some Improperly Posed Problems of Mathematical Physics. Springer, Berlin, 1967.
- MOROZOV, V.A., Methods for Solving Incorrectly Posed Problems. Springer, Berlin, 1984.
- NASHED, M.Z. (ed.), Generalized Inverses and Applications. Academic Press, New York, 1976.
- TIKHONOV, A.N. and ARSENIN, V.Y., Solutions of Ill-Posed Problems. Winson/Wiley, Washington, 1977.

c) ***Tomography*** (6 lectures in English).
Prof. Alberto F. GRÜNBAUM (Universiy of California, Berkeley).

d) ***Numerical Treatment of Ill-Posed Problems.*** (6 lectures in English).
Prof. Frank NATTERER (Univ. Münster).

Ill-posed problems are problems whose solutions are not well defined in the sense that they do not exist at all, or that they are not unique, or that they do not depend continuously on the data.

The session starts with a brief introduction into the theory of ill-posed problems, the practical difficulties, and typical areas where such problems arise.

We then present basic regularization techniques, such as truncated singular value decomposition, Tikhonov regularization, iteration, coarse discretization, and (constrained) optimization. The numerical implementation of the first two of these techniques is discussed in detail. A crucial point in both techniques is the determination of a "good" parameter controlling the trade-off between a-priori-knowledge and information contained in the data.

At the end of the session we present numerical results for selected problems in tomography.

References

- TIKHONOV, A.N. - ARSENIN, V.Y., Solution of Ill-Posed Problems. Wiley 1977.
- RAMM, A., Theory and application of some new classes of integral equations. Springer 1980.
- BERTERO, M., Problemi lineari non ben posti e metodi di regolarizzazione. Consiglio Nazionale delle Ricerche, Firenze 1982.
- LAVRENT'EV, M.M., ROMANOV, V.G., SISATSKIJ, S.P., Problemi non ben posti in Fisica matematica e Analisi, Consiglio Nazionale delle Ricerche, Firenze 1983.
- GROETSCH, C.W., The theory of Tikhonov regularization for Fredholm equation of the first kind. Pitman 1984.

Seminars

A number of seminars and special lectures will be offered during the Session.

Applications

Those who wish to attend the Session should fill in an application form and mail it to the Director of the Fondazione C.I.M.E. at the address below, ***not later than April 30, 1986***.

An important consideration in the acceptance of applications is the scientific relevance of the Session to the field of interest of the applicant.

Applicants are requested, therefore, to submit, along with their application, a scientific curriculum and a letter of recommendation.

Participation will only be allowed to persons who have applied in due time and have had their application accepted.

FONDAZIONE C.I.M.E.
CENTRO INTERNAZIONALE MATEMATICO ESTIVO
INTERNATIONAL MATHEMATICAL SUMMER CENTER

"Mathematical Economics"

is the subject of the Second 1986 C.I.M.E. Session.

The Session, sponsored by the Consiglio Nazionale delle Ricerche and the Ministero della Pubblica Istruzione, will take place under the scientific direction of Proff. ANTONIO AMBROSETTI (Università di Venezia), FRANCO GORI (Università di Firenze), ROBERTO LUCCHETTI (Università di Milano) at Villa «La Querceta», Montecatini Terme (Pistoia), Italy, ***from June 25 to July 3, 1986.***

Courses

a) ***Variational Problems arising from Mathematical Economics.*** (6 lectures in English).
Prof. Ivar EKELAND (CEREMADE, Univ. Paris IX).

The lectures will focus on two topics: 1. Infinite horizon optimization problems arising from Ramsey models, 2. Generalization of Hamilton-Jacobi theory arising from the theory of incentive.

Prerequisites: 1. Some familiarity with convex optimization is advisable. 2. I. Ekeland-R. Temam, Analyse convexe et problèmes variationnels, Dunod 1974.

b) ***Differentiability Techniques in the Theory of General Economic Equilibrium.*** (6 lectures in English).
Prof. Andreu MAS-COLELL (Harvard University).

Outline of content:

Applications of Calculus and Differential Topology methods to the problem of existence, optimality and local uniqueness of price equilibria will be reviewed. Comparative statics will also be considered. Both static and sequential models of economic equilibrium will be discussed

Basic references

- E. DIERKER, Topological Methods in Walrasian Economics. Lecture Notes in Economics and Mathematical Sciences, 92, 1974. Springer Verlag.
- S. SMALE, Global Analysis and Economics. Ch. 8 in "Handbook of Mathematical Economics", vol. II, 1981, ed. K. Arrow and M. Intriligator, pp. 331-370. North-Holland.
- A. MAS-COLELL, The theory of general economic equilibrium: A differentiable approach. Cambridge University Press, 1985.

Prerequisites

The economics will be selfcontained althought it would be helpful to have some familiarity with a graduate textbook such as: H. Varian, Microeconomics (Norton) or E. Malinvaud, Microeconomic theory (North-Holland).

For the mathematics, advanced Calculus is all that will be really needed. However, it would be convenient if the basics of real analysis could be taken for granted and, also, if students had become familiar with the simple facts of transversality theory (as gotten for example from: J. Milnor, Topology from the Differentiable View-point, 1965, University Press of Virginia, or the first half of: Guillemin, V. and A. Pollak, Differential Topology, NJ, Prentice Hall).

c) ***Dynamical General Equilibrium Models.*** (6 lectures in English).
Prof. Jose A. SCHEINKMAN (University of Chicago).

This course will deal with dynamic general equilibrium economic models, i.e., the integration of general equilibrium analysis and dynamics. Special attention will be placed on the implications of these models to the phenomena of macroeconomic fluctuations. In order to avoid excessive prerequisites we will start with a short survey of the relevant aspects of general equilibrium theory and the theory of economic growth.

Contents

— Arrow-Debreu Theory.
— Models of Economic Growth under Uncertainty as Dynamic Equilibrium Models.
— Incomplete Market Models and Economic Fluctuations.

Main References

- ARROW, K., "The Role of Securities in the Optimal Allocation of Risk Bearing", in "Essay in the Theory of Risk-Bearing", Markham Press.
- BROCK, W.A. and MIRMAN, L.J., Optimal Economic Growth and Uncertainty: The Discounted Case, Journal of Economic Theory, 4 (1972), 233-240.
- LUCAS, ROBERT E., Asset Prices in an Exchange Economy, Econometrica, 46 (1978), 1429-1445.
- LUCAS, ROBERT E., Models of Business Cycles, (mimeo) 1985.
- SAMUELSON, P., An exact Consumption-Loan Model of Interest ...., Journal of Political Economy, 66 (1958), 67-82.
- SCHEINKMAN, JOSE, General Equilibrium Models of Economic Fluctuations: A Survey Theory, (mimeo) 1984.
- SCHEINKMAN, J. and WEISS, L., Borrowing Constraints and Aggregate Economic Activity, Econometrica, January 1986.

d) ***Topics in Noncooperative Game Theory.*** (6 lectures in English).
Prof. Shmuel ZAMIR (Hebrew Univ. Jerusalem).

— Basic concepts and Results: Games in Extensive form; Games in Strategic form; Nash Equilibria; Zero-Sum Games - Minimax Theorems.
— Multistage Games: Repeated Games; Stochastic Games; Blackwell's Approachability Nation.
— Incomplete Information: The Universal Beliefs Space; Common Knowledge; Consistency of beliefs.
— Zero-Sum Repeated Games with Incomplete Information: Minimax Maxmin and Asymptotic Value; Speed of Convergence, the Variation of a Martingale and the Normal Distribution; Games without Recursive Structure.

Basic References

- OWEN, G., 'Game Theory', Second Edition, Academic Press, 1982.
- BURGER, E., 'Introduction to the Theory of Games', Prentice-Hall, 1963.
- VON NEUMAN, L. and O. MORGENSTERN, 'Theory of Games and Economic Behavior', Princeton University Press, 1944, 1947.
- PARTHASARATHY, T. and T.E.S. RAGAVAN, 'Some Topics in Two-Person Games', American Elsevier, 1971.
- SORIN, S., 'An introduction to Two-Person Zero-Sum Repeated Games with Incomplete Information', Technical Report No. 312, 1980, Institute for Mathematical Studies in the Social Sciences, Stanford University.
- MERTENS, J.F. and S. ZAMIR, 'Formulation of Bayesian Analysis for Games with Incomplete Information', International Journal of Game Theory 14 (1985), 1-29.

Seminars

A number of seminars and special lectures will be offered during the Session.

Applications

Those who wish to attend the Session should fill in an application form and mail it to the Director of the Fondazione C.I.M.E. at the address below, ***not later than April 30, 1986.***

An important consideration in the acceptance of applications is the scientific relevance of the Session to the field of interest of the applicant.

Applicants are requested, therefore, to submit, along with their application, a scientific curriculum and a letter of recommendation.

Participation will only be allowed to persons who have applied in due time and have had their application accepted.

FONDAZIONE C.I.M.E.
CENTRO INTERNAZIONALE MATEMATICO ESTIVO
INTERNATIONAL MATHEMATICAL SUMMER CENTER

"Combinatorial Optimization"

is the subject of the Third 1986 C.I.M.E. Session.

The Session, sponsored by the Consiglio Nazionale delle Ricerche and the Ministero della Pubblica Istruzione, will take place under the scientific direction of Prof. BRUNO SIMEONE (Università di Roma "La Sapienza") at the Centro di Cultura Scientifica A. Volta, «Villa Olmo», Como (Italy), ***from August 25 to September 2, 1986.***

Program outline

The CIME Summer School on "Combinatorial Optimization" aims to present recent results and current trends in this area.

Combinatorial Optimization has a peculiar location in the map of Applied Mathematics, being placed in an interzone in the middle of Combinatorics, Computer Science and Operations Research. From a mathematical point of view, it draws on pure combinatorics, including graphs and matroids, on Boolean algebras and switching functions, partially ordered sets, group theory, linear algebra convex geometry and probability theory, as well as other tools.

Over the past years, a substantial amount of research has been devoted to the connections between Combinatorial Optimization and theoretical Computer Science, and in particular to computational complexity and algorithmic issues. Quite often, the study of combinatorial optimization problems is motivated by real-life applications, such as scheduling, assignment, location, distribution, routing, districting, design and other Operations Research applications.

Although references to actual applications will be frequently given, the emphasis of the School is on theoretical aspects of Combinatorial Optimization. Special attention will be devoted to polyhedral combinatorics and its connections with combinatorial duality theories and min-max identities; to the study of important classes of functions (either real- or binary-valued) defined on the binary n-cube, and of their significance in 0-1 optimization; to the role of submodularity (a discrete analogue of convexity) in Combinatorial Optimization; to the deep link between greedy algorithms and finite geometries such as matroids and greedoids. As a matter of fact, the interplay between structural and algorithmic properties is one of the main themes addressed by the School.

Courses

a) ***Truth Functions and Set Functions.*** (6 lectures in English).
Prof. Peter L. HAMMER (Rutgers University, USA).

Truth functions and Boolean expressions. Set functions and pseudo-Boolean expressions. Quadratic Boolean expressions. Consistency of quadratic Boolean equations and the Konig-Egervary property of graphs. The parametric form of the general solution of a quadratic Boolean equation.

Quadratic pseudo-Boolean expressions. Maximization of pseudo-Boolean expressions; reduction to the quadratic case. The conflict graphs of a set function. Quadratic graphs and the problem of their recognition.

Linear majorants of quadratic pseudo-Boolean expressions. Roofs and the height of a quadratic function. Optimal roofs and the complement of a quadratic function. Equivalence of quadratic optimization and of the weighted stable set problem in a graph.

Roof duality. Strong and weak persistence.

The gap between the height and the maximum of a quadratic function. Supermodular functions are gap-free. Gap-free functions and the weighted Konig-Egervary property of ghaphs.

Open problems.

b) ***Binary Group Polyhedra, Binary Matroids, and the Chinese Postman Problem.*** (6 Lectures in English).
Prof. Ellis L. JOHNSON (IBM Thomas J. Watson Research Center, USA).

Abstract

The area to be covered has been worked on from different directions by several people. Whitney introduced matroids and defined binary matroids and graphic matroids.

Tutte characterized graphic matroids. Lehman carried matroid idea into the Shannon switching game and did fundamental work on clutters or ports of matroids. He also developed a framework of various equivalent properties for pairs of dual clutters.

Fulkerson showed that there are then pairs of polyhedra with and elegant and simple duality between their facets and vertices. Independently, Gomory developed a theory of group problems for integer programming with some special results for binary group problems. The polyhedron for a special case, the Chinese postman problem, was characterized by Edmonds and Johnson and seen to be an instance of this general framework.

Seymour further characterized binary clutters and those for which a strong integrality property, the max-flow min-cut property, holds for binary clutters. The emphasis here will be on polyhedra, optimization problems, and the binary group approach.

A dual, or blocking, framework is developed in which relations between various results can be seen.

References

- A. LEHMAN, A Solution of the Shannon Switching Game, SIAM J. App. Math. 12 (1984), 687-725.
- A. LEHMAN, On the Width-Lenght Inequality, Math. Prog. 17 (1979), 403-417.
- R.E. GOMORY, Some Polyhedra Related to Combinatorial Optimization, Lin. Algebra and Applications 2 (1969), 451-558.
- D.R. FULKERSON, Blocking Polyhedra, in 'Graph Theory and Its Applicantions', B. Harris (ed.), Academic Press, N.Y., 1970, 93-112.
- J. EDMONDS and E.L. JOHNSON, Matchings, Euler Tours, and the Chinese Postman Problem, Math. Prog. 5 (1973), 88-124.
- P.D. SEYMOUR, Matroids with the Max-flow Min-cut Property, J. of Comb. Theory B 23 (1977), 189-222.
- G. GASTOU and E.L. JOHNSON, Binary Group and Chinese Postman Polyhedra, to appear in Math. Prog..

c) ***Algorithmic Principles and Combinatorial Structures.*** (6 lectures in English).
Prof. Bernhard KORTE (Universität Bonn).

Abstract

This series of lectures discusses several algorithmic principles in combinatorial optimization.

One goal of this survey will be the study of the relations between algorithmic approaches and the underlying combinatorial structure. A combinatorial optimization problem can be defined either on a simple structure, but then it needs complicated and sophisticated algorithmic tools. Or it can be defined on a combinatorial structure which is mathematically substantial. Then it needs an easy algorithm (e.g. greedy approach). These trade-offs between structure and algorithms are studied in the first part of the lectures.

The second part is basically related to the greedy algorithm and its variants.

We discuss greedy approaches from an algorithmic as well as from a structural point of view.

Finally, we study greedoids which were introduced as relaxations of matroids by L. Lovasz and the lecturer. It turns out that many "real world" combinatorial optimization problems and structures can be formulated in the framework of greedoids.

We discuss polyhedral, structural and algorithmic aspects of greedoids. Very many special greedoids will be investigated, among which antimatroid have a considerably rich structure.

References

a) General introduction into the subject

- WELSH, D.J.A., Matroid Theory, Academic Press, London, 1976.
- LAWLER, E.L., Combinatorial Optimization: Networks and Matroids. Holt, Reinhard & Wiston, New York, 1976.
- PAPADIMITRIOU, C.H. and STEILITZ, K., Combinatorial Optimization: Algorithms and Complexity. Prentice-Hall, Englewood Cliffs, 1982.

b) Special literature
- KORTE, B. and LOVASZ, L., Polymatroid greedoids. J. Comb. Theory (series B) 38 (1985), 41-72.
- KORTE, B. and LOVASZ, L., Greedoids and linear objections functions. SIAM J. Algebraic Discrete Methods 5 (1984), 229-238.
- KORTE, B. and LOVASZ, L., Shelling structures, convexity and a happy end. in: 'Graph Theory and Combinatorics', Proceedings of the Cambridge Combinatorial Conference in Honour of Paul Erdős, Academic Press, London, 1984, pp. 219-232.
- KORTE, B. and LOVASZ, L., Structural properties of greedoids, Combinatorica 3 (1983), 359-374.
- KORTE, B. and LOVASZ, L., Greedoids - a structural framework for the greedy algorithm. in W.R. Pulleyblank (ed.): Progress in Combinatorial Optimization. Proceedings of the Silver Jubilee Conference on Combinatorics. Waterloo, June 1982, Academic Press, London, pp. 221-243

d) ***Submodular Functions, Graphs and Optimization.*** (6 lectures in English).
Prof. Eugene L. LAWLER (University of California, Berkeley).

— Matroids and submodular functions.
— Classical Network flow theory: a review.
— Duality theorems via the Max-flow Min-cut Theorem of polymatroidal network flows.
— The matroid parity problem.
— Linear time algorithms for optimal subgraph problems.
— Solution of combinatorial optimization problems by distributed computation.
— Recent results in scheduling theory.

References

- E.L. LAWLER, Combinatorial Optimization: Networks and Matroids, Holt Reinhart and Wiston, 1976.
- E.L. LAWLER and C.U. MARTEL, Flow Network Formulations od Polymatroidal Optimization Problems, Annals Discrete Math., 16 (1982), 189-200.
- E.L. LAWLER and C.U. MARTEL, Computing Maximal 'Polymatroidal' Network Flows, Math. of Operations Research, 7 (1982), 334-347.
- E.L. LAWLER, A Fully Polynomial Approximation Scheme for the Total Tardiness Problem, Operations Research Letters, 1 (1982), 207-208.
- E.L. LAWLER, Preemptive Scheduling of a Single Machine to Minimize the Sum of Completion Times, to appear in Operations Research Letters.
- E.L. LAWLER, Recent Results in Machine Scheduling Theory, Mathematical programming: The State of the Art, A. Bachem et al., Springer Verlag, 1983, pp. 202-234.
- E.L. LAWLER, P. TONG and V. VAZIRANI, Solving the Weighted Parity problem for Gammoids by Reduction to Graphic Matching, Progress in Combinatorial Optimization, W.R. Pulleyblank, ed., Academic Press, 1984, pp. 363-374.
- E.L. LAWLER, Submodular Functions and Polymatroid Optimization, in 'Combinatorial Opimization: Annotated Bibliographies', M. O'hEigerartaigh, J.K. Lenstra, A.H.G. Rinnooy Kan, Eds., J. Wiley, 1985.
- E.L. LAWLER, Why Certain Subgraph Computations Require Only Linear Time, Proc. 26th Annual IEEE Symposium on Foundations of Computer Science, October 1985.

Seminars

A number of seminars and special lectures will be offered during the Session.

Applications

Those who wish to attend the Session should fill in an application form and mail it to the Director of the Fondazione C.I.M.E. at the address below, ***not later than May 15, 1986.***

An important consideration in the acceptance of applications is the scientific relevance of the Session to the field of interest of the applicant.

Applicants are requested, therefore, to submit, along with their application, a scientific curriculum and a letter of recommendation.

Participation will only be allowed to persons who have applied in due time and have had their application accepted.

LIST OF C.I.M.E. SEMINARS

			Publisher
1954 -	1.	Analisi funzionale	C.I.M.E.
	2.	Quadratura delle superficie e questioni connesse	"
	3.	Equazioni differenziali non lineari	"
1955 -	4.	Teorema di Riemann-Roch e questioni connesse	"
	5.	Teoria dei numeri	"
	6.	Topologia	"
	7.	Teorie non linearizzate in elasticità, idrodinamica, aerodinamica	"
	8.	Geometria proiettivo-differenziale	"
1956 -	9.	Equazioni alle derivate parziali a caratteristiche reali	"
	10.	Propagazione delle onde elettromagnetiche	"
	11.	Teoria della funzioni di più variabili complesse e delle funzioni automorfe	"
1957 -	12.	Geometria aritmetica e algebrica (2 vol.)	"
	13.	Integrali singolari e questioni connesse	"
	14.	Teoria della turbolenza (2 vol.)	"
1958 -	15.	Vedute e problemi attuali in relatività generale	"
	16.	Problemi di geometria differenziale in grande	"
	17.	Il principio di minimo e le sue applicazioni alle equazioni funzionali	"
1959 -	18.	Induzione e statistica	"
	19.	Teoria algebrica dei meccanismi automatici (2 vol.)	"
	20.	Gruppi, anelli di Lie e teoria della coomologia	"
1960 -	21.	Sistemi dinamici e teoremi ergodici	"
	22.	Forme differenziali e loro integrali	"
1961 -	23.	Geometria del calcolo delle variazioni (2 vol.)	"
	24.	Teoria delle distribuzioni	"
	25.	Onde superficiali	"
1962 -	26.	Topologia differenziale	"
	27.	Autovalori e autosoluzioni	"
	28.	Magnetofluidodinamica	"
1963 -	29.	Equazioni differenziali astratte	"
	30.	Funzioni e varietà complesse	"
	31.	Proprietà di media e teoremi di confronto in Fisica Matematica	"

1964 -	32. Relatività generale	C.I.M.E.
	33. Dinamica dei gas rarefatti	"
	34. Alcune questioni di analisi numerica	"
	35. Equazioni differenziali non lineari	"
1965 -	36. Non-linear continuum theories	"
	37. Some aspects of ring theory	"
	38. Mathematical optimization in economics	"
1966 -	39. Calculus of variations	Ed. Cremonese, Firenze
	40. Economia matematica	"
	41. Classi caratteristiche e questioni connesse	"
	42. Some aspects of diffusion theory	"
1967 -	43. Modern questions of celestial mechanics	"
	44. Numerical analysis of partial differential equations	"
	45. Geometry of homogeneous bounded domains	"
1968 -	46. Controllability and observability	"
	47. Pseudo-differential operators	"
	48. Aspects of mathematical logic	"
1969 -	49. Potential theory	"
	50. Non-linear continuum theories in mechanics and physics and their applications	"
	51. Questions of algebraic varieties	"
1970 -	52. Relativistic fluid dynamics	"
	53. Theory of group representations and Fourier analysis	"
	54. Functional equations and inequalities	"
	55. Problems in non-linear analysis	"
1971 -	56. Stereodynamics	"
	57. Constructive aspects of functional analysis (2 vol.)	"
	58. Categories and commutative algebra	"
1972 -	59. Non-linear mechanics	"
	60. Finite geometric structures and their applications	"
	61. Geometric measure theory and minimal surfaces	"
1973 -	62. Complex analysis	"
	63. New variational techniques in mathematical physics	"
	64. Spectral analysis	"

1974 -	65. Stability problems		Ed. Cremonese, Firenze
	66. Singularities of analytic spaces		"
	67. Eigenvalues of non linear problems		"
1975 -	68. Theoretical computer sciences		"
	69. Model theory and applications		"
	70. Differential operators and manifolds		"
1976 -	71. Statistical Mechanics		Ed. Liguori, Napoli
	72. Hyperbolicity		"
	73. Differential topology		"
1977 -	74. Materials with memory		"
	75. Pseudodifferential operators with applications		"
	76. Algebraic surfaces		"
1978 -	77. Stochastic differential equations		"
	78. Dynamical systems		Ed. Liguori, Napoli and Birkhäuser Verlag
1979 -	79. Recursion theory and computational complexity		Ed. Liguori, Napoli
	80. Mathematics of biology		"
1980 -	81. Wave propagation		"
	82. Harmonic analysis and group representations		"
	83. Matroid theory and its applications		"
1981 -	84. Kinetic Theories and the Boltzmann Equation	(LNM 1048)	Springer-Verlag
	85. Algebraic Threefolds	(LNM 947)	"
	86. Nonlinear Filtering and Stochastic Control	(LNM 972)	"
1982 -	87. Invariant Theory	(LNM 996)	"
	88. Thermodynamics and Constitutive Equations	(LN Physics 228)	"
	89. Fluid Dynamics	(LNM 1047)	"
1983 -	90. Complete Intersections	(LNM 1092)	"
	91. Bifurcation Theory and Applications	(LNM 1057)	"
	92. Numerical Methods in Fluid Dynamics	(LNM 1127)	"
1984	93. Harmonic Mappings and Minimal Immersions	(LNM 1161)	"
	94. Schrödinger Operators	(LNM 1159)	"
	95. Buildings and the Geometry of Diagrams	(LNM 1181)	"
1985 -	96. Probability and Analysis	(LNM 1206)	"
	97. Some Problems in Nonlinear Diffusion	(LNM 1224)	"
	98. Theory of Moduli	to appear	"

Note: Volumes 1 to 38 are out of print. A few copies of volumes 23,28,31,32,33,34,36,38 are available on request from C.I.M.E.

Vol. 1062: J. Jost, Harmonic Maps Between Surfaces. X, 133 pages. 1984.

Vol. 1063: Orienting Polymers. Proceedings, 1983. Edited by J.L. Ericksen. VII, 166 pages. 1984.

Vol. 1064: Probability Measures on Groups VII. Proceedings, 1983. Edited by H. Heyer. X, 588 pages. 1984.

Vol. 1065: A. Cuyt, Padé Approximants for Operators: Theory and Applications. IX, 138 pages. 1984.

Vol. 1066: Numerical Analysis. Proceedings, 1983. Edited by D.F. Griffiths. XI, 275 pages. 1984.

Vol. 1067: Yasuo Okuyama, Absolute Summability of Fourier Series and Orthogonal Series. VI, 118 pages. 1984.

Vol. 1068: Number Theory, Noordwijkerhout 1983. Proceedings. Edited by H. Jager. V, 296 pages. 1984.

Vol. 1069: M. Kreck, Bordism of Diffeomorphisms and Related Topics. III, 144 pages. 1984.

Vol. 1070: Interpolation Spaces and Allied Topics in Analysis. Proceedings, 1983. Edited by M. Cwikel and J. Peetre. III, 239 pages. 1984.

Vol. 1071: Padé Approximation and its Applications, Bad Honnef 1983. Prodeedings. Edited by H. Werner and H.J. Bünger. VI, 264 pages. 1984.

Vol. 1072: F. Rothe, Global Solutions of Reaction-Diffusion Systems. V, 216 pages. 1984.

Vol. 1073: Graph Theory, Singapore 1983. Proceedings. Edited by K.M. Koh and H.P. Yap. XIII, 335 pages. 1984.

Vol. 1074: E.W. Stredulinsky, Weighted Inequalities and Degenerate Elliptic Partial Differential Equations. III, 143 pages. 1984.

Vol. 1075: H. Majima, Asymptotic Analysis for Integrable Connections with Irregular Singular Points. IX, 159 pages. 1984.

Vol. 1076: Infinite-Dimensional Systems. Proceedings, 1983. Edited by F. Kappel and W. Schappacher. VII, 278 pages. 1984.

Vol. 1077: Lie Group Representations III. Proceedings, 1982–1983. Edited by R. Herb, R. Johnson, R. Lipsman, J. Rosenberg. XI, 454 pages. 1984.

Vol. 1078: A.J.E.M. Janssen, P. van der Steen, Integration Theory. V, 224 pages. 1984.

Vol. 1079: W. Ruppert. Compact Semitopological Semigroups: An Intrinsic Theory. V, 260 pages. 1984

Vol. 1080: Probability Theory on Vector Spaces III. Proceedings, 1983. Edited by D. Szynal and A. Weron. V, 373 pages. 1984.

Vol. 1081: D. Benson, Modular Representation Theory: New Trends and Methods. XI, 231 pages. 1984.

Vol. 1082: C.-G. Schmidt, Arithmetik Abelscher Varietäten mit komplexer Multiplikation. X, 96 Seiten. 1984.

Vol. 1083: D. Bump, Automorphic Forms on GL (3,IR). XI, 184 pages. 1984.

Vol. 1084: D. Kletzing, Structure and Representations of Q-Groups. VI, 290 pages. 1984.

Vol. 1085: G.K. Immink, Asymptotics of Analytic Difference Equations. V, 134 pages. 1984.

Vol. 1086: Sensitivity of Functionals with Applications to Engineering Sciences. Proceedings, 1983. Edited by V. Komkov. V, 130 pages. 1984

Vol. 1087: W. Narkiewicz, Uniform Distribution of Sequences of Integers in Residue Classes. VIII, 125 pages. 1984.

Vol. 1088: A.V. Kakosyan, L.B. Klebanov, J.A. Melamed, Characterization of Distributions by the Method of Intensively Monotone Operators. X, 175 pages. 1984.

Vol. 1089: Measure Theory, Oberwolfach 1983. Proceedings. Edited by D. Kölzow and D. Maharam-Stone. XIII, 327 pages. 1984.

Vol. 1090: Differential Geometry of Submanifolds. Proceedings, 1984. Edited by K. Kenmotsu. VI, 132 pages. 1984.

Vol. 1091: Multifunctions and Integrands. Proceedings, 1983. Edited by G. Salinetti. V, 234 pages. 1984.

Vol. 1092: Complete Intersections. Seminar, 1983. Edited by S. Greco and R. Strano. VII, 299 pages. 1984.

Vol. 1093: A. Prestel, Lectures on Formally Real Fields. XI, 125 pages. 1984.

Vol. 1094: Analyse Complexe. Proceedings, 1983. Edité par E. Amar, R. Gay et Nguyen Thanh Van. IX, 184 pages. 1984.

Vol. 1095: Stochastic Analysis and Applications. Proceedings, 1983. Edited by A. Truman and D. Williams. V, 199 pages. 1984.

Vol. 1096: Théorie du Potentiel. Proceedings, 1983. Edité par G. Mokobodzki et D. Pinchon. IX, 601 pages. 1984.

Vol. 1097: R.M. Dudley, H. Kunita, F. Ledrappier, École d'Éte de Probabilités de Saint-Flour XII – 1982. Edité par P.L. Hennequin. X, 396 pages. 1984.

Vol. 1098: Groups – Korea 1983. Proceedings. Edited by A.C. Kim and B.H. Neumann. VII, 183 pages. 1984.

Vol. 1099: C.M. Ringel, Tame Algebras and Integral Quadratic Forms. XIII, 376 pages. 1984.

Vol. 1100: V. Ivrii, Precise Spectral Asymptotics for Elliptic Operators Acting in Fiberings over Manifolds with Boundary. V, 237 pages. 1984.

Vol. 1101: V. Cossart, J. Giraud, U. Orbanz, Resolution of Surface Singularities. Seminar. VII, 132 pages. 1984.

Vol. 1102: A. Verona, Stratified Mappings – Structure and Triangulability. IX, 160 pages. 1984.

Vol. 1103: Models and Sets. Proceedings, Logic Colloquium, 1983, Part I. Edited by G.H. Müller and M.M. Richter. VIII, 484 pages. 1984.

Vol. 1104: Computation and Proof Theory. Proceedings, Logic Colloquium, 1983, Part II. Edited by M.M. Richter, E. Börger, W. Oberschelp, B. Schinzel and W. Thomas. VIII, 475 pages. 1984.

Vol. 1105: Rational Approximation and Interpolation. Proceedings, 1983. Edited by P.R. Graves-Morris, E.B. Saff and R.S. Varga. XII, 528 pages. 1984.

Vol. 1106: C.T. Chong, Techniques of Admissible Recursion Theory. IX, 214 pages. 1984.

Vol. 1107: Nonlinear Analysis and Optimization. Proceedings, 1982. Edited by C. Vinti. V, 224 pages. 1984.

Vol. 1108: Global Analysis – Studies and Applications I. Edited by Yu.G. Borisovich and Yu.E. Gliklikh. V, 301 pages. 1984.

Vol. 1109: Stochastic Aspects of Classical and Quantum Systems. Proceedings, 1983. Edited by S. Albeverio, P. Combe and M. Sirugue-Collin. IX, 227 pages. 1985.

Vol. 1110: R. Jajte, Strong Limit Theorems in Non-Commutative Probability. VI, 152 pages. 1985.

Vol. 1111: Arbeitstagung Bonn 1984. Proceedings. Edited by F. Hirzebruch, J. Schwermer and S. Suter. V, 481 pages. 1985.

Vol. 1112: Products of Conjugacy Classes in Groups. Edited by Z. Arad and M. Herzog. V, 244 pages. 1985.

Vol. 1113: P. Antosik, C. Swartz, Matrix Methods in Analysis. IV, 114 pages. 1985.

Vol. 1114: Zahlentheoretische Analysis. Seminar. Herausgegeben von E. Hlawka. V, 157 Seiten. 1985.

Vol. 1115: J. Moulin Ollagnier, Ergodic Theory and Statistical Mechanics. VI, 147 pages. 1985.

Vol. 1116: S. Stolz, Hochzusammenhängende Mannigfaltigkeiten und ihre Ränder. XXIII, 134 Seiten. 1985.

Vol. 1117: D.J. Aldous, J.A. Ibragimov, J. Jacod, Ecole d'Été de Probabilités de Saint-Flour XIII – 1983. Édité par P.L. Hennequin. IX, 409 pages. 1985.

Vol. 1118: Grossissements de filtrations: exemples et applications. Seminaire, 1982/83. Edité par Th. Jeulin et M. Yor. V, 315 pages. 1985.

Vol. 1119: Recent Mathematical Methods in Dynamic Programming. Proceedings, 1984. Edited by I. Capuzzo Dolcetta, W.H. Fleming and T. Zolezzi. VI, 202 pages. 1985.

Vol. 1120: K. Jarosz, Perturbations of Banach Algebras. V, 118 pages. 1985.

Vol. 1121: Singularities and Constructive Methods for Their Treatment. Proceedings, 1983. Edited by P. Grisvard, W. Wendland and J.R. Whiteman. IX, 346 pages. 1985.

Vol. 1122: Number Theory. Proceedings, 1984. Edited by K. Alladi. VII, 217 pages. 1985.

Vol. 1123: Séminaire de Probabilités XIX 1983/84. Proceedings. Edité par J. Azéma et M. Yor. IV, 504 pages. 1985.

Vol. 1124: Algebraic Geometry, Sitges (Barcelona) 1983. Proceedings. Edited by E. Casas-Alvero, G.E. Welters and S. Xambó-Descamps. XI, 416 pages. 1985.

Vol. 1125: Dynamical Systems and Bifurcations. Proceedings, 1984. Edited by B.L.J. Braaksma, H.W. Broer and F. Takens. V, 129 pages. 1985.

Vol. 1126: Algebraic and Geometric Topology. Proceedings, 1983. Edited by A. Ranicki, N. Levitt and F. Quinn. V, 423 pages. 1985.

Vol. 1127: Numerical Methods in Fluid Dynamics. Seminar. Edited by F. Brezzi, VII, 333 pages. 1985.

Vol. 1128: J. Elschner, Singular Ordinary Differential Operators and Pseudodifferential Equations. 200 pages. 1985.

Vol. 1129: Numerical Analysis, Lancaster 1984. Proceedings. Edited by P.R. Turner. XIV, 179 pages. 1985.

Vol. 1130: Methods in Mathematical Logic. Proceedings, 1983. Edited by C.A. Di Prisco. VII, 407 pages. 1985.

Vol. 1131: K. Sundaresan, S. Swaminathan, Geometry and Nonlinear Analysis in Banach Spaces. III, 116 pages. 1985.

Vol. 1132: Operator Algebras and their Connections with Topology and Ergodic Theory. Proceedings, 1983. Edited by H. Araki, C.C. Moore, Ş. Strătilă and C. Voiculescu. VI, 594 pages. 1985.

Vol. 1133: K.C. Kiwiel, Methods of Descent for Nondifferentiable Optimization. VI, 362 pages. 1985.

Vol. 1134: G.P. Galdi, S. Rionero, Weighted Energy Methods in Fluid Dynamics and Elasticity. VII, 126 pages. 1985.

Vol. 1135: Number Theory, New York 1983–84. Seminar. Edited by D.V. Chudnovsky, G.V. Chudnovsky, H. Cohn and M.B. Nathanson. V, 283 pages. 1985.

Vol. 1136: Quantum Probability and Applications II. Proceedings, 1984. Edited by L. Accardi and W. von Waldenfels. VI, 534 pages. 1985.

Vol. 1137: Xiao G., Surfaces fibrées en courbes de genre deux. IX, 103 pages. 1985.

Vol. 1138: A. Ocneanu, Actions of Discrete Amenable Groups on von Neumann Algebras. V, 115 pages. 1985.

Vol. 1139: Differential Geometric Methods in Mathematical Physics. Proceedings, 1983. Edited by H. D. Doebner and J. D. Hennig. VI, 337 pages. 1985.

Vol. 1140: S. Donkin, Rational Representations of Algebraic Groups. VII, 254 pages. 1985.

Vol. 1141: Recursion Theory Week. Proceedings, 1984. Edited by H.-D. Ebbinghaus, G.H. Müller and G.E. Sacks. IX, 418 pages. 1985.

Vol. 1142: Orders and their Applications. Proceedings, 1984. Edited by I. Reiner and K. W. Roggenkamp. X, 306 pages. 1985.

Vol. 1143: A. Krieg, Modular Forms on Half-Spaces of Quaternions. XIII, 203 pages. 1985.

Vol. 1144: Knot Theory and Manifolds. Proceedings, 1983. Edited by D. Rolfsen. V, 163 pages. 1985.

Vol. 1145: G. Winkler, Choquet Order and Simplices. VI, 143 pag 1985.

Vol. 1146: Séminaire d'Algèbre Paul Dubreil et Marie-Paule Mallia Proceedings, 1983–1984. Edité par M.-P. Malliavin. IV, 420 pag 1985.

Vol. 1147: M. Wschebor, Surfaces Aléatoires. VII, 111 pages. 1985

Vol. 1148: Mark A. Kon, Probability Distributions in Quantum Statisti Mechanics. V, 121 pages. 1985.

Vol. 1149: Universal Algebra and Lattice Theory. Proceedings, 19 Edited by S. D. Comer. VI, 282 pages. 1985.

Vol. 1150: B. Kawohl, Rearrangements and Convexity of Level Sets PDE. V, 136 pages. 1985.

Vol 1151: Ordinary and Partial Differential Equations. Proceeding 1984. Edited by B.D. Sleeman and R.J. Jarvis. XIV, 357 pages. 19

Vol. 1152: H. Widom, Asymptotic Expansions for Pseudodifferen Operators on Bounded Domains. V, 150 pages. 1985.

Vol. 1153: Probability in Banach Spaces V. Proceedings, 1984. Edit by A. Beck, R. Dudley, M. Hahn, J. Kuelbs and M. Marcus. VI, 4 pages. 1985.

Vol. 1154: D.S. Naidu, A.K. Rao, Singular Pertubation Analysis Discrete Control Systems. IX, 195 pages. 1985.

Vol. 1155: Stability Problems for Stochastic Models. Proceeding 1984. Edited by V.V. Kalashnikov and V.M. Zolotarev. VI, 44 pages. 1985.

Vol. 1156: Global Differential Geometry and Global Analysis 1984 Proceedings, 1984. Edited by D. Ferus, R.B. Gardner, S. Helgaso and U. Simon. V, 339 pages. 1985.

Vol. 1157: H. Levine, Classifying Immersions into $\mathbb{R}^4$ over Stable Map of 3-Manifolds into $\mathbb{R}^2$. V, 163 pages. 1985.

Vol. 1158: Stochastic Processes – Mathematics and Physics. Proceedings, 1984. Edited by S. Albeverio, Ph. Blanchard and L. Streit. VI 230 pages. 1986.

Vol. 1159: Schrödinger Operators, Como 1984. Seminar. Edited by S. Graffi. VIII, 272 pages. 1986.

Vol. 1160: J.-C. van der Meer, The Hamiltonian Hopf Bifurcation. VI, 115 pages. 1985.

Vol. 1161: Harmonic Mappings and Minimal Immersions, Montecatini 1984. Seminar. Edited by E. Giusti. VII, 285 pages. 1985.

Vol. 1162: S.J.L. van Eijndhoven, J. de Graaf, Trajectory Spaces, Generalized Functions and Unbounded Operators. IV, 272 pages. 1985.

Vol. 1163: Iteration Theory and its Functional Equations. Proceedings, 1984. Edited by R. Liedl, L. Reich and Gy. Targonski. VIII, 231 pages. 1985.

Vol. 1164: M. Meschiari, J.H. Rawnsley, S. Salamon, Geometry Seminar "Luigi Bianchi" II – 1984. Edited by E. Vesentini. VI, 224 pages. 1985.

Vol. 1165: Seminar on Deformations. Proceedings, 1982/84. Edited by J. Ławrynowicz. IX, 331 pages. 1985.

Vol. 1166: Banach Spaces. Proceedings, 1984. Edited by N. Kalton and E. Saab. VI, 199 pages. 1985.

Vol. 1167: Geometry and Topology. Proceedings, 1983–84. Edited by J. Alexander and J. Harer. VI, 292 pages. 1985.

Vol. 1168: S.S. Agaian, Hadamard Matrices and their Applications. III, 227 pages. 1985.

Vol. 1169: W.A. Light, E.W. Cheney, Approximation Theory in Tensor Product Spaces. VII, 157 pages. 1985.

Vol. 1170: B.S. Thomson, Real Functions. VII, 229 pages. 1985.

Vol. 1171: Polynômes Orthogonaux et Applications. Proceedings, 1984. Edité par C. Brezinski, A. Draux, A.P. Magnus, P. Maroni et A. Ronveaux. XXXVII, 584 pages. 1985.

Vol. 1172: Algebraic Topology, Göttingen 1984. Proceedings. Edited by L. Smith. VI, 209 pages. 1985.